# Analyzing Census Microdata

# Analyzing Census Microdata

## Angela Dale, Ed Fieldhouse
Cathie Marsh Centre for Census and Survey Research,
University of Manchester, UK

### and

## Clare Holdsworth
Geography Department, University of Liverpool, UK

**With contributions from**
Paul Boyle, University of St Andrews, UK
Thomas Cooke, University of Connecticut, USA
Karen Glaser, King's College London, UK
Myles Gould, University of Leeds, UK
Emily Grundy, London School of Hygiene and Tropical Medicine, UK
Keith Halfacree, University of Wales Swansea, UK
Kelvyn Jones, University of Portsmouth, UK
Suzanne Model, University of Massachusetts, USA
Mike Murphy, London School of Economics, UK
Stephen Simpson, Cathie Marsh Centre for Census and Survey Research, UK
Darren Smith, University of Leeds, UK

**And extra material from**
Mark Brown, Cathie Marsh Centre for Census and Survey Research, UK
Roger Jones, The Australian National University, Australia
Chuck Humphrey, University of Alberta, Canada
Phil Murphy, University of Swansea, UK
Mark Tranmer, Cathie Marsh Centre for Census and Survey Research, UK

John Wiley & Sons, Ltd

First published in Great Britain in 2000 by
Arnold, a member of the Hodder Headline Group,
338 Euston Road, London NW1 3BH
©2000 Angela Dale, Ed Fieldhouse and Clare Holdsworth

John Wiley & Sons Ltd, The Atrium, Southern Gate, Chichester, West Sussex, PO19 8SQ, United Kingdom

*British Library Cataloguing in Publication Data*
A catalogue record for this book is available from the British Library

*Library of Congress Cataloging-in-Publication Data*
A catalog record for this book is available from the Library of Congress

ISBN 13: 978-0-047-68919-6

12345678910

Typeset in 10/12 pt Times by Academic + Technical, Bristol

# Biographical information about the authors and contributors

**Paul Boyle** is Professor of Human Geography at the University of St Andrews, Scotland. He has interests in migration and health, is on the executive editorial board of the *International Journal of International Geography* and is co-editor of the *Population and Migration Studies* Routledge series. He is co-author of *Exploring Contemporary Migration* (Longman, 1998) and co-editor of *Migration into Rural Areas* (Wiley, 1998) and *Migration and Gender in the Developed World* (Routledge, 1999). A further co-edited book, *The Geography of Health Inequalities in the Developed World* (Ashgate), and a singly authored textbook, *Migration and Health* (Routledge), are due to be published in 2000.

**Thomas J. Cooke** is an Associate Professor of Geography at the University of Connecticut, where he is also the Director of Urban Studies. His research focuses on how the interrelationships between social mobility and spatial mobility are mediated by gender, race, and ethnicity. This research focus is demonstrated in two broad areas. First, he is concerned with whether the continuing suburbanization of job opportunities in US cities negatively impacts the employment prospects of segregated ethnic and racial minorities. Second, he is concerned with how women's employment prospects are limited by family migration decision-making prospects.

**Angela Dale** is Director of the Cathie Marsh Centre for Census and Survey Research and Professor of Quantitative Social Research. She is co-author of *Doing Secondary Analysis* (Unwin Hyman) and co-editor of *The 1991 Census User's Guide* (HMSO) and *Analyzing Social and Political Change* (Sage).

**Ed Fieldhouse** is Deputy Director of the Cathie Marsh Centre for Census and Survey Research. He is a social scientist with an interest in social, economic and geographical inequalities, labour market behaviour and policies, and electoral politics.

**Karen Glaser** has a BA in Psychology from the University of Michigan, an MA in Latin American Studies from the University of Texas, an MSc in Demography from the London School of Economics and a PhD in Sociology from the University of Michigan. She joined the staff of the Age Concern Institute of Gerontology at King's College London as a researcher in 1994 and was subsequently made a lecturer.

**Myles Gould** has a BA in Geography from Portsmouth Polytechnic and a PhD from the University of Bristol. He was appointed Research Associate at Portsmouth University in 1993 and then in 1996 Lecturer in Health Services Research at the NHS S&W funded R&D Support Unit. In 1998 he was appointed Lecturer in Geography at the University of Leeds where he teaches Census Data and Analysis Methods, Quantitative Research Methods and the Geography of Health. He has recently co-authored *Epidemiology: An Introduction* (Open University Press).

**Emily Grundy,** after a first degree in history, studied Medical Demography in the Centre for Population Studies at the London School of Hygiene and Tropical Medicine, which she now heads. In between she has worked at the University of Nottingham Medical School, the City University and King's College London. Her main interests are in family and household demography and the health and mortality of older populations. She has published widely in both areas.

**Keith Halfacree** has a BSc in Geography from Bristol University and a PhD from Lancaster University. He joined the Department of Geography at the University of Wales, Swansea in 1991. He has authored and co-authored numerous academic papers on rural studies, population migration and political geography. Recently he co-wrote (with Paul Boyle and Vaughan Robinson) *Exploring Contemporary Migration* (Longman, 1998) and co-edited (with Boyle) *Migration into Rural Areas: Theories and Issues* (Wiley, 1998) and *Migration and Gender in the Developed World* (Routledge, 1999).

**Clare Holdsworth** is a Lecturer in the Department of Geography, University of Liverpool. She is a demographer with a particular interest in household and family formation.

**Kelvyn Jones** graduated with first-class honours BSc in Geography from the University of Southampton and subsequently obtained a PhD from that University in 1980. He is currently Professor and Head of the Department of Geography at the University of Portsmouth and is active in the University's Institute for the Geography of Health. His main academic interests are in the analysis of data with complex structure and the effects of place characteristics on health and illness. He has taught an advanced modelling course at the Essex European Summer School throughout the 1990s.

**Suzanne Model** holds a PhD in Social Work and Sociology from the University of Michigan. Since 1985, she has been a member of the Department of Sociology at the University of Massachusetts in Amherst. Her research focuses on the socio-economic attainment of ethnic minorities, with special emphasis on cross-national comparisons. Over the past five years, she has held visiting positions at the University of Manchester, the Institut du Longitudinal in Paris and Erasmus University in Rotterdam.

**Mike Murphy** has degrees in Mathematics, from Cambridge University, and in Quantum Mechanics and General Relativity, from York University. He spent 10 years working as a Statistician in the Central Statistical Office, London, and two years as a Research Fellow at the London School of Hygiene and Tropical Medicine before moving to the London School of Economics and Political Science in 1980, where he is Professor of Demography and Convener of the Department of Social Policy.

**Stephen Ludi Simpson** is currently employed by Bradford Council and the Cathie Marsh Centre for Census and Survey Research, UK. He has been responsible for committee and editorial work with the Local Authorities Research and Intelligence Association, the journal *Statistics in Society,* the Central and Local Government Information Partnership and the Radical Statistics Group. His interests lie in demography and the use of statistics by communities and governments.

**Darren Smith** is Teaching Fellow in the School of Geography at the University of Leeds. His research interests include gendered dimensions and labour market consequences of family migration, processes of sociocultural, physical and economic change in urban and rural locations. He is currently co-editing *Contemporary Geographies of Leeds* (University of Leeds Press).

# Preface

In 1990, the UK Census Offices agreed to release Samples of Anonymised Records from the 1991 Census of Population in response to a request from the Economic and Social Research Council. Although samples of microdata from the census had been available in the US, Canada, Australia and some European countries – as long ago as the early 1960s for the US – this was a first for the UK. It marked a major step forwards in providing a new form of census output and a new resource for academic and policy-related research.

The ESRC set up a unit specifically to support and disseminate the SARs – the Census Microdata Unit at the University of Manchester. The unit's first director was the late Professor Cathie Marsh who had played a major role in making the scientific case for the release of the SARs. Tragically, Cathie died of cancer before the SARs were released by the Census Offices. This book is dedicated to the memory of Cathie and her achievements in securing census microdata for the UK.

In 1996, the Census Microdata Unit organized a conference at Manchester to highlight the specific contribution of census microdata to social science research. The papers presented highlighted features such as the large sample size that allows analysis of small subgroups, the relatively detailed geography, the wide population coverage, including those living in institutions, and the scope for international comparison. This book focuses on the potential of census microdata as distinct from both aggregate census output and also microdata from social surveys. In addition, it includes, in Chapter 8, some exemplars of analyses based on contributions to the conference.

The book is designed to meet the needs of those familiar with census data but unfamiliar with analyzing microdata and also those familiar with survey data but new to the census. Thus Chapters 1 and 2 highlight the unique aspects of the census and the strengths of samples of microdata drawn from censuses. Chapter 3 provides an introduction to the scope for making international comparisons using census microdata. Chapter 4 gives essential guidance on how to select relevant populations, the various levels of analysis that can be used, the likely sources of error in the data and strategies for assessing the precision of estimates. In Chapter 5 we provide step-by-step guidance in deriving both relatively simple and more complex variables and in Chapter 6, for those not familiar with analyzing microdata, we discuss the basic exploratory analysis necessary before proceeding to more sophisticated analysis.

Chapter 7 is concerned with multivariate regression analysis (OLS and logistic) and provides an introduction to multilevel modelling. Finally, Chapter 8 brings together a set of exemplars that highlight some of the research strengths of census microdata.

We hope that the book will be a timely tool for those wishing to analyze the 2000 round of censuses. All the census producers referred to in the book have excellent web sites that provide full and up-to-date information on the methodology and content of their census. We have not, therefore, attempted to include detailed information for each country but, rather, a framework for understanding the overall structure and content and the main differences between countries.

# Acknowledgements

The book has benefited from the knowledge of contributors and commentators from around the world. We are grateful to Chuck Humphrey, University of Alberta, for commenting on the Canadian material, and supplying valuable additional material and to Roger Jones, Australian National University, for providing comparable help with Australian material. We are also grateful to Mark Ellis for helpful comments on the US census. We give particular thanks to Mark Brown, CCSR, who provided invaluable assistance in tracking down information from statistical offices around the world in order to complete Table 2.1 and in providing much other help and support.

Thanks are also due to Nicki Dennis from Edward Arnold who commissioned the book and chivvied us to meet deadlines, to Liz Gooster, who succeeded her, and to Kirsty Stroud and Anke Ueberberg.

Much of the work on the book has been supported under the ESRC/JISC 1991 Census programme, grant number H507255140.

The SARs were provided through the Census Microdata Unit of the University of Manchester, with the support of the ESRC/JISC/DENI. All tables containing SAR data, and the results of analysis, are reproduced with the permission of the Controller of Her Majesty's Stationery Office (Crown Copyright).

The US PUMS were provided by the Inter-university Consortium for Political and Social Research, Ann Arbor and the ESRC Data Archive. The data were produced by the U.S. Dept. of Commerce, Bureau of the Census, Washington, DC.

Angela Dale and Ed Fieldhouse, CCSR, University of Manchester and
Clare Holdsworth, Department of Geography, University of Liverpool.
March 2000

# 1

# An introduction to census data

## 1.1 Introduction

In this chapter we provide a framework in which to locate census data generally and microdata files more specifically. We aim to identify those features which distinguish census data from other social survey data and the implications of this for analysis. In so doing we highlight differences between countries and try to identify the basis for these differences.

Most countries in the world conduct regular censuses of their population and for many countries this is a statutory requirement, either at national or at supranational (e.g. European Union) level. As such, the public has a legal obligation to comply, and failure to do so may result in penalties. This gives the census a status not accorded to other national surveys, whether conducted for government, business or academia. It also brings with it an obligation by government to protect the data supplied by the public and to ensure that confidentiality pledges are not broken.

Despite statutory requirements, census-taking can only be effective with the consent of the population. Ensuring the public acceptability of the census is taken very seriously by national statistical institutes (NSIs)[1] and has implications that run through all aspects of the census-taking process, from data collection to data release policies. It influences the number of questions asked, the type of questions, the procedures used for processing and storage of data, the outputs generated and the conditions under which data are made available outside the NSI. In some European countries this consent is no longer present and census-taking has broken down – for example, Germany has not taken a full population census since the census planned for 1983 had to be postponed until 1987 because of public concern over proposals to use census returns to update the local population registers (Redfern, 1987). The Netherlands has not taken a census since 1971, following a significant level of refusal in the 1971 Census and poor tests results in 1979 (Redfern, 1987). Both countries rely on alternative data sources, particularly population and housing

---

[1] We are using the term 'national statistical institute' to include national statistical offices (such as the Office for National Statistics in the UK ) and agencies such as the US Census Bureau and the Australian Bureau of Statistics.

registers and sample surveys. In Denmark also, use of register-based information has replaced conventional census-taking.

The value of administrative sources in supplementing census data is increasingly being recognized, particularly as technological advances make it a cost-effective alternative which reduces respondent burden. For example, Norway and Sweden continue taking a census whilst also making extensive use of register-based information for statistical purposes (Redfern, 1987). One of the aims of the decentralization of statistics in Italy and the formation of the National Institute of Statistics (ISTAT) was to exploit the opportunity to make better use of administrative sources for statistical purposes (Garonna, 1994). In the UK, administrative sources played an important role in identifying the under-count in the 1991 Census and will also be used as supplementary information in the 2001 Census (Dale, 1993; Brown *et al.*, 1999). In a consultation by Statistics Canada on the 2001 Census, the most frequent comment on income data was that it should be collected by file linkage with Revenue Canada (Lathe, 1995). This would reduce the response burden of the census and, if adopted, respondents could be asked for permission to collect information on tax rates, thereby giving a better idea of purchasing. The feasibility of this is being studied for future censuses.

In most countries, then, a population census remains of central importance in providing core data, but with increasing recognition of the value of register-based information as a supplement and, in some cases, a substitute.

The self-completion nature of the census also has wide-ranging implications for data collection. It is immediately apparent that the prospect of conducting an entire population census by interview – for all but the smallest and richest of countries – is completely out of the question. The questions asked in a census, and the way in which they are organized, must therefore be clear and comprehensible to the entire population, irrespective of age, ethnicity, level of education or regional variation. There is no interviewer to weave through complex filters, or to ensure that questions have been correctly understood, or even to encourage the respondent to take part. While in many countries enumerators may provide some assistance with problems in form-filling and also conduct some checks on completion, this is at a very different level from the input that would be made by a highly skilled and experienced social survey interviewer.

## 1.2  Why are census data important?

One of the most important roles played by census data is in providing an accurate count of the population for small areas, which can be used by government as a basis for determining political representation or in allocating resources to local government.

In the USA, a decennial census is required by the Constitution in order to determine how many members of the House of Representatives each of the 50 states is entitled to. This is done through an 'apportionment' process that divides up the 435 seats amongst the 50 states. Other major uses are to define the geographical boundaries of state legislative districts and to allocate nearly $200 billion in annual federal funding. In Canada, the first post-Confederation census, held in 1871, was also designed to determine appropriate representation by population in the new parliament and is still the basis for the allocation of seats for each province in the House of Commons (Statistics Canada, 1992a). Census data are also important for setting the boundaries of Canadian federal electoral districts and determining transfer payments to the provinces and territories. The provinces also use the census for apportionment of seats in their legislatures and in determining resource allocation to municipalities. In addition, the official languages of Canada, which are protected under the Canadian Charter of Rights and Freedoms, require the census to collect data about mother tongue, that is, the language first learned as a child at home and that is still understood.

In Australia the census also provides a basis for estimating the population for each state, territory and local government area, mainly for electoral purposes and the distribution of government funding. In Britain, baseline information about the population is used by central government for resource allocation to local government and health authorities. For example, census-based indicators are key components in a formula to calculate spending allocations for local authorities in the form of a standard spending assessment (SSA). In France and Spain, census data are also important in providing population counts at a local level, with resource allocation directly related to this.

The fact that these data are collected and used by government for purposes of national accounting and resource allocation gives them a unique status and authority. It also means that access to the data by those outside central government becomes very important, often in order to challenge decisions made by government, e.g. by demonstrating that a different formula may result in different spending allocations.

The census also represents a (near) complete count of the population, capturing place of residence, housing and social and economic characteristics at a single point in time. This has several consequences. Firstly, data are available for small areas without any sampling error. They are also available on a comparable basis across areas. This unique aspect of census data is of considerable importance in policy analysis (e.g., to determine the demographic profile of an area; to make comparisons about the relative needs of areas; to assess which areas have the greatest housing need or the poorest health). The value of this information operates at various levels. National governments need census data to assess the distribution of resources to the next tier of government (e.g., states or local authorities); state, local government or health authorities may use these data to target resources within their area.

In most countries the census is also unique in including the institutional population; typically, social surveys are confined to the 'private household' population. For many analysis purposes this sector of the population is routinely omitted, but there are some specific instances where their inclusion is important. For example, research on the health of elderly people may be biased and can lead to misinterpretation if those living in institutions are omitted (Glaser et al., 1997b, and Ch. 8). Analyses of residential mobility are also valuable in showing the flow of elderly people into residential accommodation and how this varies with marital status and household composition (Glaser and Grundy, 1998).

## 1.3 What influences the content and timing of the census?

### The legal requirement

We have already indicated that census data are of great policy significance and that policy requirements are a major influence on the range and nature of the questions asked. The wide range of uses by central and local government (e.g., household projections, planning health services, monitoring the impact of race relations legislation) ensures that topic coverage extends beyond basic demographic information (e.g. age, sex, household composition), to include housing, employment, education and, for a number of countries, income questions.

In most countries there is primary legislation that requires a census to be taken and sets the overall parameters. We have already seen that in the USA, a decennial census is required under the Constitution. In Britain the 1920 Census Act gave the Registrars General for England and Wales and Scotland the general authority and duty to conduct a census at intervals no shorter then every 5 years; reports on a census have to be laid before Parliament. However, the recent UK devolution legislation means that responsibility for census-taking in Scotland, as for Northern Ireland, has been devolved. In Canada the British North America Act of 1867 established the decennial census which was then replaced by a series of acts culminating in the Statistics Act of 1970, requiring a census of the population to be taken every fifth year. In Australia, the Census and Statistics Act 1905 provided that 'the census shall be taken in the year 1911, and in every 10th year thereafter', although the timing was disrupted by the economic depression and World War II. Since 1961, a census has been taken every 5 years, a practice that is now mandatory under an amendment to the Act in 1977 (ABS, 1996).

### The role of supranational bodies

The content of the census may also be influenced by supranational considerations – most importantly the United Nations and, for European countries, the European Union. The Statistical Office of the United Nations makes recommendations on a set of 'basic' topics, including demographic variables, employment and education, and housing, that should be included in the decennial round of censuses. The European Union, through its statistical office, EUROSTAT, requires data on a core set of topics to be supplied by each EU member state. In some cases, a country is able to obtain a derogation (e.g. the age of the household's dwelling is a required question, but Britain has successfully argued that this information cannot be collected by a census). Other countries, as discussed earlier, compile the required population information from registers rather than a census. The UN recommend the inclusion of a further set of questions although there is considerable variation in the extent to which these are asked. Despite a UN recommendation on the inclusion of income, this was not asked by any of the EU member states in the 1990 round of censuses and Britain was the only EU member state to ask ethnic group (Langevin et al., 1992). The role of supranational bodies in cross-country comparability is discussed in Section 3.3.2.

## The need for comparability over time

The use of the census in a benchmarking capacity also gives rise to a requirement for comparability over time. Thus local government may use the census to chart the change in housing tenure or housing amenities within an area, or changes in the modes of travelling to work. This often produces a tension between retaining exactly the same questions and updating questions to reflect a change in social mores. Change may result in adding a question, dropping a question no longer relevant (e.g. indoor WC) or extending the available responses to include new categories. Whilst very small changes can sometimes have quite far-reaching implications, failing to change questions to reflect changes in conditions or behaviour can also have serious consequences. An example is given below.

Until 1991, British censuses had not recorded cohabiting relationships. In 1991, however, reflecting the rapid rise in cohabitation rates since 1981, the relationship question was changed so that 'cohabiting' was offered as a possible response to 'relationship to reference person/head of household'. There is then no comparability between cohabitation rates in 1981 and 1991 but, much more importantly, it affects the incidence of lone fathers in 1981 and highlights the consequences of failing to provide appropriate categories. In the 1981 Census, a cohabiting couple with a child would have been recorded as two single people with a child. If the father was recorded as the 'reference person', the relationship of his partner to him would have been correctly recorded as 'unrelated' whilst that of the child would have been 'son' or 'daughter'. Thus a father–child relationship would have been established but not a 'mother–child' relationship, resulting in a lone-father family and a female 'other' person in the household. Harrop and Plewis (1993) have shown, using 1981 Census microdata from the ONS Longitudinal Study,[2] that this resulted in 3.0 per cent of families with a dependent child headed by a lone father, compared with an estimate of 1.6 per cent from the General Household Survey. The 2001 Census will be the first to record same sex couples – generating lack of comparability with 1991 when same sex couples were defined as 'unrelated'.

Marital status in Canada is also undergoing a major society change. The shift amongst younger age cohorts to common law unions has been dramatic over the past decade and, in 1991, a separate question was asked to identify common law status in addition to a question on legal marital status. Prior to 1991, those living in a common law relationship were instructed to record their legal marital status: 'separated', 'divorced', 'widowed' or 'never married'.

## Acceptability to the public

We have already suggested that, even though there may be a legal requirement on the public to take part in the census, this cannot be imposed against the public will. Therefore census offices go to considerable lengths to ensure that questions are acceptable.

---

[2] The ONS Longitudinal Study links together a 1 per cent sample of individuals from the 1971 Census, with the same individuals in the 1981 and 1991 Census. For each census, a sample based on four common dates of birth is drawn and individuals matched on the basis of their National Health Service identification number and other characteristics. Information on births and deaths, drawn from vital registration records, is also linked into the study (Hattersely and Creeser, 1995).

The cultural differences between countries in what is considered acceptable give rise to some interesting differences in question content. An obvious example is income. This is included in the USA, Canada and Australia, but in Britain the question has always been deemed too sensitive to ask. British government surveys show considerable non-response on income; however, much of this is due to inadequate information rather than non-response. For example, from 1979 to 1991 about 25 per cent of respondents to the General Household Survey were classified as missing on derived variables for 'total income'. However, from 1992, procedures were changed to allow estimates and proxy information to be accepted and this reduced the level of missing information to 5–9 per cent. Nonetheless, some subgroups of the population still present difficulties – for example, about 12 per cent of the self-employed have missing income data (Thomas, 1994). An income question was included in the 1997 Census Test in Britain. At the time of publication of the White Paper on the 2001 Census (Cmnd 4253) in March 1999, a decision over including the question had not been made. The White Paper discusses the need for income to be collected in 2001 against the difficulties of collecting accurate information and concerns over public disquiet and possible consequences for the risk to the census as a whole. A final decision not to ask income in the census was made early in 2000. This was based on the adequacy of alternative data sources and, in particular, the ability to impute small area level data from social survey data.

Questions on race or ethnicity provide another example where there is considerable variation between countries in acceptability. For example, the topic would fall outside the legal framework in France which does not allow distinctions to be made on the basis of race or ethnic group (Coleman and Salt, 1996) and, as we saw above, Britain is distinctive within the European Union in asking a census question on ethnic group. In those countries which include a question on race or ethnic group in their census, there are considerable differences in the timing of the introduction of the question (as far back as 1870 in the USA) and in the rationale for its inclusion. This is discussed in detail for the USA, Canada, Australia and Britain in Section 3.5.

## Cost and response burden

The cost of data collection is strongly related to the response burden on the public, and both of these factors also influence the length and complexity of the census form; again, there is considerable variation between countries, but also variation over time. In the USA and Canada, two different forms are used: a short form and a long form. The latter goes to a sample of households (about 17 per cent in the USA and 20 per cent in Canada) and contains a more extensive set of questions than the short form. For example, the 2000 short form for the USA asks only seven questions and takes about 10 minutes to complete, whilst the long form has an additional 27 questions and takes on average 38 minutes to complete. The 1991 Canadian short form asked only nine questions – reduced to seven in 1996 – whilst an additional 44 – 48 in 1996 – were included on the long form. However, in both the USA and Canada, public sector cuts have, at various times, resulted in serious consideration being given to restricting the census to the short form only. So far this has not been necessary.

In Britain and Australia, the same form is sent to the entire population and therefore there is no opportunity to ask extra questions of a sample. The number of questions is typically greater than included in a 'short' form but less than in a 'long' form: in the 1981 British Census, 21 questions were asked and this increased to 25 in 1991 and several additional questions are planned for 2001; the 1991 and 1996 Australian Censuses included 47 questions. Field sampling was tried unsuccessfully in Britain in the 1961 Census and has never been repeated. The reason for the failure was that the sampling was influenced by the enumerators who were responsible for delivering a long form at every 10th household. Countries which use address lists are able to draw a sample before a list is issued to an enumerator – thereby avoiding the problem of enumerator bias. The alternative to field sampling, used in Britain in recent decades, has been to draw a 10 per cent sample of households from the enumerated population and to restrict coding of complex questions such as occupation to this sample. However, it has been decided that a sample will not be taken from the 2001 Census for full coding. Instead, greater use will be made of technological developments, such as optical character recognition and automated coding, to reduce costs sufficiently to enable coding of the entire population.

## The self-completion nature of the census

We have already pointed out that the very size of a complete enumeration requires it to be self-completion. (In Britain, this has been the case since 1841!) This has immediate consequences, not just for the number of questions that can be asked but also for the level of complexity. It means that there can be very little filtering and that questions must be straightforward, without the need for complex instructions. They must also be understandable and meaningful to all sections of society and use words which have the same meaning to the different groups who will answer them. This is a daunting task and is exacerbated in countries with very big differences in culture, language and lifestyle, either between different regions or between ethnic or social groups. In Australia the remoteness of many Aboriginal communities and their special characteristics have led to the development of an Indigenous Enumeration Strategy to inform indigenous communities about the census; data for discrete localities are obtained by interview rather than self-enumeration (Martin and Taylor, 1995). In Canada, where language differences create the need for separate census forms in French and English, respondents have the choice as to which form they wish to complete. British census forms are provided in English or Welsh in Wales. For other non-English speaking communities, explanatory leaflets are available in a range of different languages, but the forms themselves are not translated.

Although there is some tailoring of the way in which census data are collected to meet the needs of specific communities, in general the range of requirements that have to be met by a census means that there is a considerable amount of comparability between countries in the topics covered, if not the actual question wording. The strong policy influence on the census leads to a high degree of continuity in questions. Limitations on the number of questions which can be asked also means that very strong arguments are needed for a new question to be included and questions are unlikely to be introduced to reflect very immediate and short-term concerns.

## 1.4 The availability of census data for analysis

Demands for more data from the census to be released continue unrelentingly, fuelled by an increased ability to process and analyze the data. It has been argued that census data are very influential in the policy process, particularly in terms of resource allocation. There is a very strong argument that the democratic process demands that these data should be available for re-analysis by those outside government – particularly as the costs of census-taking are borne by the public purse (Dale, 1998). (Although many countries charge for census products such as tables or microdata files, income from these sources comes nowhere near the full costs of a census and often reflects only the marginal cost of production.) Thus the availability of census data for further analysis plays a very important role in allowing challenges to interpretation by government, or different perspectives to be taken using the same data.[3]

In the USA this role is explicitly recognized, in the context of federal statistical agencies in the report of a panel on confidentiality and data access:

> ... to enhance the integrity of research findings, independent analysts should have access to data, regardless of the organisation that collected it. As a critical element in the democratic process, this access can allow reanalysis by groups with different agendas; stimulate new inquiries in important social, economic and scientific questions; lead to improvements in the quality of data through suggestions for better measurement and data collection methods; and provide information to improve government forecasts and allocations.
>
> (Duncan et al., 1993, p. 31)

In the USA the freedom of information legislation has resulted in census data being available in the public domain, free of copyright.

In Canada, also, Data Liberation (Albert et al., 1996) explicitly recognizes the mutual benefit that can accrue from academic analysis of government statistical data, whilst the British Office for National Statistics contains, in a foreword to its framework document, the following words by the then Chancellor of the Exchequer, Kenneth Clarke:

> Reliable social and economic statistics are fundamental to the Citizen's Charter and to open government. Open access to official statistics provides the citizen with both a picture of society and a window on the work and performance of government, showing the scale of government activity in every area of public policy and allowing the impact of government policies and actions to be assessed.
>
> (ONS, 1996b)

However, countering these very strong claims for access to data is the respondent's right to confidentiality of the information supplied. Whilst this is recognized as fundamentally important by all NSIs, it is particularly acute in the case of data which is collected under statute. Obvious identifiers like names and addresses are always removed from data, but then there remains the task of deciding whether

---

[3] The basis for financing the UK Census of Population means that data are sold to all sectors – even to central government. Census data for academic use have been purchased by JISC (Joint Information Systems Committee) and the ESRC (Economic and Social Research Council). Access to the 2001 Census should be dramatically improved for central government, local authorities and academia through a Treasury-funded access programme.

only tabular data should be available or whether microdata can also be released. For both aggregate and microdata releases, decisions have to be taken concerning the smallest geographical area that is 'safe' and the amount of detail available on each variable. If microdata files are released, then there are further decisions over the number of files and the sample size and structure of each, the level of geography and the detail to be released. Thus is discussed more fully in Section 2.3.

In general, though, there is a balance of interests between the amount of data that should be released and the risk to the respondent of a breach of confidentiality. For most countries, the optimal point is one where the risk to the respondent is minimal. However, this may vary in relation to the prevailing culture concerning privacy and the role of the state (which, in turn, can often be related to historical events), and the assessment of the gain from making census data available as against the consequences of a breach of confidentiality. In the final analysis, census offices have a paramount concern to retain the public's confidence as, without this, further census-taking would become impossible.

## 1.5 Range of datasets produced from the census

One may visualize the availability of census data on a continuum from, at the most restrictive end, summary tables with very limited detail on either individuals, households or areas to, at the other extreme, a microdata file with complete population coverage and high level of detail on all variables. Very few, if any, countries lie at either extreme and, in general, it is possible to trace a development over time along this continuum. Changes over time have been assisted by technological developments, although the extent to which technological innovations have been implemented is related to the perceived balance of interest between ease of access to data and the increased risk to confidentiality that this may bring.

In most countries, the minimum output from the census is determined by statutory requirements, e.g. to produce basic demographic data at a small area level to meet needs associated with the allocation of resources or with electoral representation. In Britain, the 1920 Census Act lays down the outputs that are legally required (Section 4 (1)) and then refers to discretionary powers for other outputs. Similarly, in the USA there is a basic constitutional requirement to meet the needs of apportionment, but many other aspects are discretionary.

Early censuses produced printed reports with tables for areas and topics. With the advent of computers (in Britain the 1961 Census was the first to use a computer) came the possibility of producing more tabulations, held in electronic form, and relatively easily transportable across sites. At the same time there was a big increase in the demand for information from the census, as computers allowed faster processing of the data. In Britain this meant an increase in the local statistics output from 28 tables and 1571 counts in 1971 to 82 tables and 8722 counts in 1991. Generally there is an established trade-off, set within country-specific parameters, between size of sample and detail available, where detail includes both geographic and individual/household information. Thus tables with a complete population count, at very low levels of geography, have the least amount of individual information. Using an example from Britain, the lowest geographical area in the 1991 Census had on average about 200 households and up to four variables in a cross-tabulation. By contrast, microdata files provide very considerable individual detail – typically holding 40–60 variables for each person, many at quite detailed levels. As a trade-off for this detail they have a heavily restricted sample size (typically between 1 and 5 per cent) and much less geographical detail, usually not going below areas of 100 000 population.

In subsequent chapters we shall identify the specific analysis potential of census microdata, but in the following section we consider the problems and pitfalls of using census data in general, irrespective of the form in which it is used.

## 1.6 Methodological issues related to the census

We have suggested above that census data are of considerable analytic value and this book is aimed at promoting the use of census microdata. It is therefore important to assess the quality of census data and, in particular, to assess any limitations and short-coming. Chapter 2 will provide a further assessment of methodological issues specific to microdata.

The role of the census in providing baseline data for resource allocation and for political representation means that great importance is attached to the accuracy of the population count and to the characteristics of those missed by the census. Therefore most countries conduct methodological work immediately after a census to assess *coverage* (the extent to which the population has been accurately counted) and *quality* (the accuracy of answers to the questions in the census). It is unusual to find this level of checking in sample surveys. Although the methods used vary between country and also change between censuses, the availability of this coverage assessment is a major benefit for the analyst.

### 1.6.1 Coverage of census data

The census aims for a complete enumeration of households and people. However, in many western countries there has been a rise in concern over census response rates. The 1991 Census for Britain as a whole missed about 2.2 per cent of the population (1.3 million residents) and a further 1.6 per cent were included through records based on imputation (OPCS, 1994). These coverage results are comparable to those for Canada (an estimated under-enumeration of just under 3 per cent of the population in 1991), the USA and Australia (both just under 2 per cent in 1990). Wiggins (1993) and Kerr (1998) both provide reviews of recent methods of detecting the undercount in these countries.

In Britain there have been post-enumeration checks conducted after each census since 1971 and they have become an essential part of the census process (Dale, 1993). Methodological work following the 1991 Census to estimate the extent of under-enumeration and its characteristics in Britain (Heady *et al.*, 1994) revealed the importance of drawing on a range of independent data sources in order to make adequate checks. Of particular concern in Britain has been the realization that the extent of under-enumeration varies very much between geographical areas and between different sociodemographic groups. For example, census output for highly urbanized areas is estimated to have missed 20–25 per cent of men in their 20s (OPCS/GRO(S), 1994). Where unadjusted estimates are used for resource allocation then these areas – often the areas of greatest need – will be underfunded by comparison with areas with lower levels of under-enumeration (Simpson and Dorling, 1994). Alternative estimates, drawing on a range of data sources in addition to the census, have been calculated for all wards in Britain in 1991 (Simpson *et al.*, 1997).

In the Australian 1991 census, the highest rates of under-enumeration were for Aboriginal and Torres Strait Islanders (over 5 per cent), the Northern Territory (4.1 per cent) and Western Australian (2.1 per cent) populations. Other under-enumeration figures are: people not at their usual address on census night (14.1 per cent); persons born in New Zealand (3.5 per cent); those living outside metropolitan areas (2.2 per cent); those aged 15–19 (2.5 per cent); 20–24 (3.6 per cent); and 25–29 (3.4 per cent) with males undercounted more than females (ABS, 1995). In Canada, the highest under-enumeration is for both men and women aged 20–24 and 25–29, with relatively low levels of under-enumeration amongst children.

In the USA, the undercount of 1.8 per cent for 1990 masks considerable variation between population subgroups, notably by race. For example, the undercount for blacks was 5.7 per cent, compared with 1.3 per cent for non-blacks. Demographic analysis of the 1990 Census suggested an overcount for those aged 15–19 and 20–24, which contrasts sharply with the UK, Canada and Australia, where young people have the highest levels of under-enumeration (Kerr, 1998).

The difficulties experienced with the 1991 UK Census have led the Census Offices to a growing recognition of the need for an estimate of under-enumeration, independent of the census. A Census Coverage Survey (CCS) has been planned for 3–4 weeks after the 2001 Census in the UK. This will be an interview survey with about 300 000 households, designed to measure the level of undercount and also the characteristics of the households and individuals missed by the census. The census and the CCS will be compared and exact matches identified using a combination of automated and clerical matching. Unlike earlier censuses, quality of answers to census questions will be assessed separately as part of the census rehearsal, held in 2000.

The US Census Bureau will conduct a similar exercise – the Accuracy and Coverage Evaluation (ACE) – following the 2000 Census. The ACE sample will consist of about 300 000 housing units drawn from an independent listing of housing in the blocks chosen for the sample. These housing units will be matched to census housing units for the same blocks and the resulting final list used to interview people at each household. The list of people interviewed will be matched to the Census 2000 results and the differences reconciled. In earlier censuses, the Bureau has also assessed under-enumeration by attempting to match, for the same geographical area, a sample of census enumerations against a sample from the population that has been drawn independently of the census. This dual-record or capture-recapture approach is described in detail by Hogan (1993).

Traditionally, under-enumeration in Canada has been estimated using a number of methods, the most important of which is a reverse record check whereby a sample of persons who should have been enumerated in the census is selected from other sources (e.g. a population listing) and then located in the census (Statistics Canada, 1992a). Thus a check is made from a population sample to the census. In Australia, a measure of the extent of census undercounting is obtained through a post-enumeration survey of a sample of households conducted shortly after the census, while data quality for various topics is assessed through consistency checks, comparisons with data from other sources and demographic analysis.

## *Adjusting for under-enumeration*

Whilst there is general agreement on the need to identify the extent of under- (or over-) enumeration in a census, there is considerable debate over how to use the information. In the USA there has been considerable controversy over whether or not to adjust the census estimates. The mayors of large cities such as New York, where under-enumeration represents a potential loss of revenue, have sued the Census Bureau for the effects of under-enumeration and have demanded adjustment (Choldin, 1994). However, adjustment would also have consequences for allocation of seats to the House of Representatives which uses a formula sensitive to small differences between state populations (Choldin, 1994). For the 2000 Census it has been ruled that the ACE will not be used to adjust the census figures for the purposes of reapportionment, although the adjusted results will be made available to government for other purposes (US Bureau of the Census website www.census. gov). However, in the UK, adjustment to the 2001 Census will be made as part of the census-processing procedures. Termed the 'One Number Census', a single database will be created for the entire population, using not only the census returns but also a process of imputation whereby households and individuals are added to the census database and their characteristics estimated using information from the CCS (see above). This database will then be used to generate all statistics from the 2001 Census which will aggregate to a single population estimate. The methodological strategy for the One Number Census is described in detail in Brown *et al.* (1999). The One Number Census represents a radically different approach to that used in the 1991 Census where, in addition to adding imputed wholly absent households, a series of other adjustments were made. These included allowances for under-enumeration of elderly people, infants, the Armed Forces and migrants from overseas, in order to meet the population estimates obtained from 'rolling forward' the 1981 Census figures (Marsh, 1993c).

In Australia, no adjustment for undercounting is made to census data, but published estimates of the resident population by age, sex and region are adjusted. Begeot *et al.* (1993) provide an overview of methods used by European countries to check the results of their census. These are summarized in Table 1.1. Most countries conduct a post-census sample survey, whilst others use comparisons with a population register or with previous census estimates. The majority of countries use more than one method and many use three.

Methods of checking the coverage and quality of census data will change and evolve over time as new problems emerge and improved methods become available. Thus in Britain the methods used for the 2001 Census are likely to differ from those used in earlier censuses. However, the key point to be made is that a great deal of time and effort is spent in identifying those sections of the population most likely to be missed by the census and those questions which are particularly prone to misinterpretation. For the analyst of census data, this methodological work is of immense value.

**Table 1.1** Summary of methods used by European countries to check the results of their census. (Reproduced from Begeot *et al.*, 1993, with permission.)

|  | Previous estimate | Population register | Post-census sample survey | Labour force survey | Other method |
|---|---|---|---|---|---|
| Belgium |  | Y |  |  | Y |
| Greece |  | Y |  |  |  |
| Spain |  | Y | Y |  |  |
| France |  |  | Y |  |  |
| Irish Republic | Y |  |  |  | Y |
| Italy |  | Y | Y |  |  |
| Luxembourg | Y | Y |  |  |  |
| Portugal | Y |  | Y |  | Y |
| United Kingdom | Y |  | Y |  | Y |
| Austria | Y |  | . |  | Y |
| Norway |  |  | Y |  |  |
| Sweden |  |  | Y | Y |  |
| Switzerland | Y |  |  | Y | Y |

## 1.6.2 Quality of census data

The quality of data captured by the census is related to the method of data collection. Almost universally, forms are self-completion and therefore there is no interviewer to check consistency or accuracy of responses. In particular, the filtering and consistency checks that are facilitated by computer-assisted interviewing (CAI) are not available. There is considerable variation in the extent to which enumerators are used to assist in the completion of forms.

In the 1991 Census in Britain and Australia, enumerators were required to collect the census forms from each household, to check that all sections of the form had been fully completed, and to provide whatever assistance was needed, particularly where the householder was elderly or had difficulties with the form (Dale, 1993; ABS, 1996). For the 2001 Census, Britain is planning to change its procedures to ask householders to post back completed census forms. Enumerators will then be used to target areas where follow-ups are needed.

The US and Canada use mail-out and mail-back for the enumeration of most areas; however, in both countries, areas considered hard to enumerate are targeted by enumerators. For the 2000 Census, the US Census Bureau plans to follow up 100 per cent of households which have not responded by the end of the period of mail response. In Canada in 1991, about 2 per cent of households were enumerated by means of an interview conducted by the census representative. This method was used in remote or northern areas, or Indian reserves where irregular mail services made mail-back impractical (Statistics Canada, 1992a).

Quality checks carried out soon after the census establish those questions which are susceptible to error. The reports from this work are placed into the public domain and are thus available to the analyst. In Britain the quality check on the 1991 Census was made by trained survey interviewers who returned to a sample of households within 3 months after the census and asked each household member to provide the same information as recorded on the census schedule (Wiggins, 1993; Heady *et al.*, 1996). This was then compared with the verbatim response given on the census schedule, and where there were discrepancies the reasons for these were explored. It is fair to say, however, that despite quality checks following the census, questions which clearly do not work still continue to be asked. Thus in Britain, a question asking the number of rooms in the household had a gross error rate of 29 per cent in the 1981 Census – almost exactly the same as that recorded in the 1991 Census (Britton and Birch, 1985; Heady *et al.*, 1996).

Generally, post-enumeration studies of the quality of census data in Britain show it to be remarkably high and fairly consistent between decades. Overall, gross errors (the extent of disagreement between the census and the interview) were less than 5 per cent, although it was higher for economic position – nearly 8 per cent for men and nearly 14 per cent for women in 1991. For women, one of the main reasons for this was that part-time working was under-recorded by the census. [In Spain, the Labour Force Survey was used to identify errors in questions on 'labour activity' in the 1991 Census results (Begeot *et al.*, 1993).] However, for other individual characteristics, such as age and educational qualifications, agreement was high, about 2–3 per cent error. Data quality was slightly lower where information had been provided by a 'form-filler' or 'head of household' on behalf of other household members. In Britain in 1991 a single form was issued to the household and one person asked to take responsibility for its completion. Heady *et al.* (1996) found, generally, that non-householders either completed their own personal information or were consulted. However, as many as 28 per cent of sons or daughters of the householder were not consulted. Levels of accuracy varied with the question and the relationship of the person to the householder. Generally, information was poorer on non-relatives if they were not consulted, although for sons and daughters there was little difference. One notable difference was on the economic activity of the spouse of the householder; if the householder did not consult, the error rate was 25 per cent, compared with 12 per cent when 'he' did consult (Heady *et al.*, 1996, p. 23).

### More information

Information on coverage and quality is generally published by NSIs as technical documents. For example, details of the Canadian programme to measure coverage error are available as technical reports from the Bureau of Statistics [e.g. *Coverage: 1991 Census Technical Reports* (Catalogue 92-341E); *Sampling and Weighting* (Catalogue 92-342E)]. These reports also describe the procedures in conducting the census. Similarly, the Australian Bureau of Statistics (1995) provides details of the undercount in the 1991 Census. Generally these documents can be found by searching the websites of the relevant NSIs; these are listed in Appendix 2.1.

## 1.6.3 The census as a cross-sectional snapshot at 5- or 10-year intervals

Like many social surveys, the census records information at a single point in time. Whilst some retrospective questions may be asked (in Britain there are questions on address 1 year ago) these are very limited and, in general, we know very little about the occupational or life-course trajectory on which an individual may be located. Research using longitudinal data (Dale and Davies, 1994) has highlighted areas where results from cross-sectional analysis may be misleading if one is not aware of the effects of inertia, sample selection bias or unobserved heterogeneity. One example comes from work on migration which is based solely on moves in the past year. Research using fuller migration histories has shown that the population can be roughly divided into 'movers' and 'stayers' – those who move frequently and those who move very little. Those who moved in the past year will contain a disproportionate number of frequent movers, who may have rather different characteristics from those who seldom move. Thus we cannot make a simple extrapolation from 1-year migration figures to those for 5 or 10 years.

Another example relates to the occupation recorded. Work-history data show that people recorded as holding the same occupation at a single point in time may be on very different occupational trajectories. In the British context, Stewart *et al.* (1980) cite the occupation of 'clerk' which, for some employees, may be a step on the way to managerial status whilst for others may represent a dead-end job. These different groups may have quite different qualifications, aspirations and lifestyle, yet cross-sectional data will obscure these differences.

For many countries (Britain, USA, Italy, Spain) there is a 10-year interval between censuses, although others (Canada, Australia) hold a 5-yearly census. In a fast-moving society a great deal may change over a 10-year period and therefore analysis based on decennial census data can soon become out of date. In Canada, even a 5-yearly census is seen as problematic, and some major urban centres conduct their own censuses because their administrators feel that the population counts for their municipalities change substantially enough between census years to warrant larger provincial grant payments than those based on the most recent Canadian census. In Britain there has been considerable pressure to provide accurate estimates at small area level on an annual basis. These are not currently produced by the Office for National Statistics, although figures are compiled on an *ad hoc* basis by commercial companies and in response to central government and academic research initiatives.

Often one of the important uses of census data is to compare change between censuses. As with all comparisons over time, considerable care is needed over changing definitions and classifications – evident from the example of lone fathers given earlier in this chapter. However, some variables, such as employment status and migration, are closely related to the state of the economy. Thus differences in levels of unemployment will depend on whether unemployment was rising or falling at the two points in time. The amount of migration in the previous year will be strongly affected by both the housing market and the labour market and, again, comparisons across a 5- or 10-year period may be quite misleading. In most countries, annual surveys (e.g. Labour Force Surveys) are available to provide estimates of change at sub-national level as well as administrative data (e.g. unemployment figures) at much smaller areas, and these are important complements to census data.

The fact that the census is a unique source of statistics for small geographical areas means that it is important to be able to establish how areas have changed between censuses. However, a comparison of geographical areas over time is also subject to difficulties caused by changes in boundaries. Without access to the geographical coordinate for each household, this often requires a great deal of work in preparing 'look-up tables' that map old areas onto new areas on a best-fit basis. In the UK there have been a number of initiatives by geographers to provide the tools for supporting these comparisons (Dorling and Atkins, 1995).

## 1.7 Technological innovations for the 2000 round of censuses

New technologies have made a considerable impact on the ease of census-taking. It is worth reviewing a range of developments across an international spectrum to provide an indication of where future developments might be going.

The growth in the availability of computerized, geo-referenced address lists should not only improve coverage but also allow the option of pre-enumeration sampling. Such lists have been used since 1970 in the USA, and for the 1990 Census a comprehensive system known as TIGER (Topological Integrated Geographical Encoding and Referencing) was created (Barr, 1993). In Britain a geo-referenced address listing is available through Ordnance Survey's Address Point and will be used to provide the initial listing of addresses in the 2001 Census. In 1991 and previous decades, enumerators in the UK had to draw up their list of addresses by walking around their district. Computerized lists can thus be issued to enumerators, or the addresses used for mailed out forms. In Canada, like the USA, computerized address lists have also been used for pre-sampling and the methodology for compiling them has recently been updated and improved.

There is a long history to the development of optical mark reading (OMR – electronic scanning of ticks or crosses) and optical character reading (OCR), and many countries, including Australia, New Zealand and the USA now use this as a speedy and accurate way to enter census data onto computer. For example, the US Census Bureau's plans for the 2000 Census include full electronic imaging and processing of questionnaires, automated sorting of mailed responses as well as optical mark recognition for check boxes and optical character recognition for write-in data. The 1996 Canadian Census used automated coding for a number of written responses that had first been alphabetically captured by direct data entry. Questions which were automatically coded included: relationship to person 1, home language, ethnic origin, place of residence one and five years ago, and place of work. These methods are being introduced in the 2001 Census in Britain and are likely to include computerized coding for occupation and industry, thereby allowing 100 per cent coding of these topics for the first time.

Technological developments also influence how census data may, in the future, be made available to users. US data from the 2000 Census will be disseminated through the internet using the *American FactFinder*. This will enable users to access prepackaged data products, documentation and on-line help, as well as to build customized data products. In the UK, 1991 aggregate small area statistics are available to academic users through web-based access at Manchester University (Harris, 2000).

Systems to provide access to census records for the entire population using on-line tabulation with software controls to limit the size and table contents (Keller-McNulty and Unger, 1998; Turton, 1999) may not be available in the short-term but represent possible modes of access in the future. However, these developments need to be assessed in terms of the benefits from greater availability of data against the risk to respondent confidentiality raised by the growing sophistication of computer hackers and the greater availability of large databases of personal information and ever-increasing power of computers. Whilst there is a growing demand for census data, this has to be balanced against the risk to public confidence that may arise if the role of NSIs in the collection of statistical information comes to be challenged by the public or by pressure groups. In Chapter 2 we return to this topic specifically with respect to microdata.

In this chapter we have provided an international context for census-taking in general. In the next chapter we look more specifically at microdata, and consider what data files are available, what influences shaped this availability, who has access to the data and on what conditions, and the ways in which the data are being used.

# 2

# An introduction to census microdata

## 2.1 Introduction

Chapter 1 provided an introduction to the census in general, with reference to a number of western countries. This chapter is specifically concerned with samples of microdata drawn from the census, i.e. records that relate to an individual, a family or a household.

Whilst a decennial census of population is usually conducted in response to a statutory requirement, the production of microdata files is discretionary. Their production is usually in response to demand from one or more sections of the user community. The structure and content of microdata files are therefore related to these user needs, the other products available, and the legislation governing data confidentiality. The production of files may also depend upon where costs lie and whether the census office has a cost recovery requirement.

These factors vary greatly between countries and also over time. Census offices may change their policy towards data release in response to public opinion or new research evidence on disclosure risk; changes in levels of public expenditure may impact upon the conduct of the census itself, as well as on the production of output products. For all of these reasons, the detailed structure, content, cost and availability of census microdata are not static but are shaped over time by a range of different influences. For example, legislation introduced by the Italian government in 1989 which decentralized statistical activities and led to the creation of a National Statistical System also led to new restrictions on the release of census microdata (Garonna, 1994). In response to user requests, Statistics Canada increased the sample size of their microdata files between 1971 and 1991 but retained the same file structure. In the USA, public expenditure cuts in the mid-1990s threatened to prevent the production of public use files from the 2000 Census. Thus legislative changes, user demand and financial constraints all influence the availability, structure and content of microdata files. Whilst there may be a general trend towards greater availability of data, the examples above show that this is not inevitable.

In this chapter we provide a general framework for the structure and content of census microdata, examine the development of microdata files in the USA, Canada, Australia and the UK, and discuss the specific contribution of microdata to the research process. The final section summarizes the basic information about microdata files for the 1990/1 and 2000/1 censuses and gives websites and other contact points for further detailed documentation.

## 2.2 The release of census microdata

### 2.2.1 An overview

The advent of computers made it possible to process, store and distribute anonymized digital data. Aggregated tabular data have been available in this form for about 50 years. These tables have generally comprised only three or four variables but usually contain a complete population count, at very small area level. By contrast, microdata files contain much more detail – perhaps 40 or 50 variables, and with fine detail on occupation, household composition or income – but only for a sample of the population and with strictly limited geographical definition. Microdata files of this kind form the focus of this chapter.

However, any table can be weighted by its cell count to provide microdata and therefore there is no hard and fast distinction between what is commonly called 'aggregate or tabular data' and 'individual or microdata' (Section 1.5). In some situations, census data are produced as very large tables with seven or eight variables and detailed geographical information. For example, the Australian Bureau of Statistics extracts standard or customized matrices or tables containing detailed cross-classified information on a number of variables for a defined population and geographic level. This may be seen as an intermediary product, lying somewhere between aggregate tabular data and microdata files with a wide range of variables on a sample of cases, often allowing linkage between individuals in families or households. In Australia, customized matrices may be ordered, specifying any combination of census variables and geographic levels, provided the matrix does not breach confidentiality rules which require a minimum of five units in each cell, although random adjustments can be applied to a limited number of small cells if necessary. These data matrices are not discussed further in this book, although it is important to recognize that there is always a balance between sample size, geographical detail and the range of other variables available.

### 2.2.2 The legal and institutional background to release of microdata

Discussion of the release of outputs from the census must, from the outset, recognize the bargain struck between the national statistical offices or agencies and the public. The legal requirement to complete the census schedule, imposed by government on the public, in turn places a heavy responsibility on the national statistical office to guarantee the confidentiality of the data provided by the public. This basic equation underpins all decisions over the release of census data. Thus in Britain census legislation makes it an offence punishable by imprisonment for any member of the census staff to disclose personal information about others without legal authority (Marsh, 1993d).

The USA led the field in the release of microdata, which is perhaps appropriate for a country with a Freedom of Information Act and where census data do not carry the copyright restrictions which operate in many other countries. The first microdata files were released from the 1960 Census – although retrospective microdata files have subsequently been extracted for earlier years (see Section 2.7). The confidentiality of the US public use microdata files (PUMS) is governed by Title 13 of the United States Code.

In Canada, release of data is governed by the 1971 Statistics Act which prohibits the publication of any personal information from which an individual could be identified. Canada first released public use microdata files (PUMFs) from the 1971 Census and have continued this policy for every quinquennial census since then. In Australia, amendments to the Census and Statistics Act in 1983 authorized the release of unidentifiable (anonymized) individual statistical records under strictly specified conditions safeguarding privacy.

In Britain the release of microdata files had been under discussion since the 1970s but was not initially accepted by the census offices because it was judged to contravene the 1920 Census Act. Under section 4.2 of this Act, requests could be made by users for the release of statistical abstracts from the census but there was uncertainty over whether samples of microdata could be considered as statistical abstracts. In the mid-1980s, Cathie Marsh led a working party of academics, set up by the Economic and Social Research Council to conduct an exhaustive analysis of the case for the release of microdata files, with a detailed assessment of the likely risk to confidentiality (Marsh et al., 1991). Following legal advice that samples of microdata could be deemed statistical abstracts, the 1991 Census White paper (Cm 430, 1988) announced that requests for abstracts in the form of anonymized records would be considered, subject to the overriding need to ensure the confidentiality of individual data (Marsh, 1993b). The case for samples of anonymized records was accepted by the Census Offices in 1989 and the structure and content of the files drawn from the 1991 Census were heavily influenced by the US and Canadian experiences.

## 2.2.3 Overview of census microdata files

Many countries have a review body which considers evidence on requests for data release and makes the final decision over what is acceptable. For example, Statistics Canada have set up a Microdata Release Committee which reviews all proposals to release microdata files. In the following paragraphs we give a brief overview of the files which are released for the USA, Canada, Australia and UK. Fuller information is contained in Appendix 2.1.

### United States of America

The US Bureau of the Census has released census microdata for every census since 1960. Three different files were released from the 1990 Census: 5 per cent and 1 per cent samples of housing units (as for 1980), and for the first time a 3 per cent sample of the elderly. The 5 per cent and 1 per cent samples have the same content and are structured so that the relationship between individuals in the same households is retained. The difference between these two files is in the geographic content of the public use microdata area (PUMA):

- The 5 per cent sample identifies all states and various subdivisions within them depending on their having a population of at least 100 000 people. Nationwide, this gives a sample of over 12 million persons and over 5 million housing units.
- The 1 per cent sample presents data for metropolitan or non-metropolitan areas or any mixture of the two, depending again on a population threshold of at least 100 000 people.

The 3 per cent sample of the elderly population contains the same geography as the 5 per cent sample.

The US PUMS contain the full range of population and housing information collected in the 1990 Census, and many of the more commonly used derived variables, e.g. household income, poverty status. The files can be purchased without any licensing procedure and are available to any organization or private individual. There is, therefore, no restriction on use outside the USA.

## Canada

Statistics Canada have produced PUMFs for every quinquennial census since 1971, retaining the same structure throughout the time period (Albert *et al.*, 1996). The PUMFs consist of three files providing national samples structured around three units of analysis:

- the individual
- the family
- the household.

Each file is rectangular, but each contains some summary contextual information about the other two units of analysis. A 1 per cent sample was released for each file from the 1971 Census; this increased to 2 per cent for the individual file in the 1980s and to 3 per cent for all files for the 1991 and 1996 Censuses.

The 3 per cent individual sample drawn from the 1991 Census contains over 119 variables for 809 654 individuals and includes sociodemographic information, place of birth, immigration, ethnic origin, language, educational qualifications, employment, occupation, industry, income, some housing information and a few summary variables about the household and family characteristics. Provinces are identified as well as census metropolitan areas with a population of at least 100 000 at the previous census.

## Australia

Microdata files were first produced for the 1981 Census, with two separate files at different geographic levels:

- a 1 per cent household file containing full classificatory detail on all person, family and dwelling variables, but without geographic variables other than a code showing section of state (major urban, other urban or rural);
- a 1 per cent sample of individuals with reduced detail on some variables but information on state or territory and capital city within mainland states.

The files from the 1986 Census contain greatly reduced detail but both include individual, family and household information with a hierarchical structure for a

**Table 2.1** Availability of census microdata in Europe

| Country | Date of census | | Microdata | | | | | |
|---|---|---|---|---|---|---|---|---|
| | Last | Next | Years available (1) | Number of files? (2) | Sample size and sampling unit (3) | Files available outside census office? (4) | Files available for researchers in other countries? (5) | Plans for 2000/2001 census (6) |
| Austria | 1991 | 2001 | Microdata are not made available outside the Statistical Office. However, Aggregated Files (of up to 30 variables) are, and can be used like microdata 1971, 1981, 1991 | (*Relates to aggregated files described previously*) 17 files Population Census: 7 Housing Census: 4 Census of Local Units of Employment: 6 | (*Relates to aggregated files described previously*) 100% | (*Relates to aggregated files described previously*) Yes, with conditions of copyright | (*Relates to aggregated files described previously*) Yes | Same concept to be continued |
| Belgium | 1991 | 2001 | 1988–1996 | Four files: Individual Housing Household + 1 smaller file (BN) from 1988 to 1996 | | Confidential files: – require authorization for specific research purpose – individuals anonymized and detailed variables removed | Yes, as in column (4) | As before |
| Bulgaria | 1992 | 2001 | 1992 | Two files: Housing (by 13 characteristics) Population (by 17 characteristics) | Population file (individuals): 8 487 317 records Housing: 3 063 149 records | Yes, after annulling the Unified Civil Register Number. Purchased commercially on CD-ROM and diskette | Yes, as in column (4) | Same files as for 1992 likely |
| Czech Republic | 1991 | 2001 | No plans to release any microdata files | – | – | – | – | – |

| | | | | | | | | |
|---|---|---|---|---|---|---|---|---|
| Denmark | 1981 | No plans | None | — | — | — | — | — |
| Finland | 1995 | 2000 | Census: 1950, 1970, 1975, 1980, 1985, 1990, 1995 (Register-based data from 1987, including 1995 Census) | Three files: Persons Families Households and Dwellings Files are in Longitudinal Census Data File | 100% (except 1950 = 10%) | Yes, subject to conditions and under discretionary powers | Yes, as in 4 | Continue with register-based census (2000/2001) |
| France | 1990 | 1999 | 1968, 1975, 1982, 1990 | Individual file | 5% sample all variables 20% sample a few variables | Yes, for researchers | Sometimes | A rolling census is under discussion |
| Germany | 1987 | 2001 | 1989, 1991, 1993, 1995 (practically anonymized primary files for the microcensus) | One for microcensus Also in progress: 'practically anonymized files' for sample survey of income and European Household Survey | 1% (microcensus) | Yes, on a contract basis | | No decision on when or under what conditions a future population census will take place |
| Greece | 1991 | 2001 | 1981, 1991 | 1981 – 1 file 1991 – 7 files | 1981, 10% 1991, 100% | Not available | Not available | Not yet |
| Iceland | 1981 | 2000 | None from census, but microdata files made available from national register of persons | | (from national register: 100%) | Yes | Yes, in principle | Not decided – files from 1981 Census may be released simultaneously with 2001 Census |
| Ireland | 1996 | 2001 | Currently none released but legislation in 1993 (1993 Statistics Act) specifies provision of microdata for research purposes | | | | | |

**Table 2.1** Continued

| Country | Date of census | | Microdata | | | | | |
| | Last | Next | Years available (1) | Number of files? (2) | Sample size and sampling unit (3) | Files available outside census office? (4) | Files available for researchers in other countries? (5) | Plans for 2000/2001 census (6) |
| --- | --- | --- | --- | --- | --- | --- | --- | --- |
| Italy | 1991 | 2001 | 1991 | 3 files: Individual Household Occupied dwellings | 1% 1% 1% | Yes, for all researchers | Yes | Similar files as for 1991 |
| Luxembourg | 1991 | 2001 | 1981, 1991 | 1 | Whole population | Yes, to Government services and research institutes working for these services | See column (4) | No details yet |
| Netherlands | No census since 1971 | No plans | None | – | – | – | – | – |
| Norway | 1990 | 2000 | 1960, 1970, 1980, 1990 | 1 | 1960: 3.6 m 1970: 3.9 m 1980: 4.1 m 1990: 4.2 m | Extracts of files made available as anonymized records. Requires licence from Norwegian Data Inspectorate | Yes in principle. Same conditions as in 4 and needs special permission from Norwegian Data Inspectorate | No details yet |
| Portugal | 1991 | 2001 | 1981, 1991 | Buildings Housing units Private households Family nuclei Resident population | Total units | No direct access. Files can be provided according to specified request | See column (4) | As stated in column (2) |

| | | | | | | | |
|---|---|---|---|---|---|---|---|
| Slovenia | 1991 | 2000 | 1981, 1991 | Four files: Individuals Households Dwellings Agricultural holdings (1991 only) | Full coverage of 'de jure' population | Yes, for research purposes only (without identification data on individuals, households or dwellings) | Yes, under same conditions as column (4) | To include same units (individuals, households, dwellings) |
| Sweden | 1990 | No plans | 1960, 1965, 1970, 1975, 1980, 1985, 1990 | Two files: Individuals Households and dwelling | 100% | Yes, after anonymization and control for data protection and data secrecy | Yes, as in column (4) | As before |
| Switzerland | 1990 | 2000 | 1970, 1980, 1990 | For individual years, 2 files: Persons Households Also a harmonized file of 1970/80/90: Persons Households Buildings/ Dwellings | Persons file 1970 = 6.3 m 1980 = 6.5 m 1990 = 7.0 m Household file 1970 = 2.1 m 1980 = 2.5 m 1990 = 3.0 m Harmonized files: Persons = 19.8 m Households = 7.6 m Buildings = 7.4 m | Yes, with signing of confidentiality contract | Only to universities in countries where law on data protection exists – a confidentiality contract must be signed | Probably individual, household and dwelling files. In addition, 2000 data will be harmonized and added to three relevant files |

1 per cent sample of households. From the 1991 Census there is a single Household Sample File (HSF) which is hierarchically structured with dwelling, family and person records from a 1 per cent sample of households and geographic levels based on regional populations of 500 000 or more. The HSF from the 1996 Census is similar to the 1991 release and the HSF from the 2001 Census is planned for release in 2003.

### The United Kingdom

Two Samples of Anonymised Records (SARs) have been extracted from the 1991 Censuses for England and Wales, Scotland and Northern Ireland:

- A 2 per cent sample of individuals in households and communal establishments. This comprises 1.1 million records containing information on all the topics asked in the census and limited information about other members of the household. This file has a minimum population size of 120 000, with sub-threshold areas grouped with neighbouring areas.
- A 1 per cent hierarchical sample of households and individuals in those households. This comprises 215 000 households and the 542 000 individuals enumerated in them. The full range of census variables is available, with region as the lowest level of geography.

The data for England, Wales and Scotland are combined in the same files whilst data for Northern Ireland are held in separate files, reflecting the slightly different schedule that was used and the different processing procedures. However, a combined file for the UK, which maximizes comparability, is available to analysts. Plans for 2001 SARs have been submitted to the Census Offices.

### Europe

A number of other European countries have released samples of microdata from their census. Table 2.1 gives a detailed description of which countries release microdata samples and the basis upon which they are available.

Websites for all European NSIs are given in Appendix 2.1 and these may give more detailed or more up-to-date information.

In all these examples, a common feature has been the confidence that the data released is 'safe', i.e. that there is no reasonable likelihood of being able to identify individuals or households. In Section 2.4 we review the methods used for making these assessments. First, we briefly mention an alternative way in which microdata may be made available – through a safe setting.

## 2.3 Census microdata: 'safe data' versus 'safe setting'

Some European countries, in particular Scandinavia, have linked census microdata to administrative records and thereby increased the amount of data available for analysis – often also adding a longitudinal dimension. These data files typically match records for sampled individuals from census to census, adding in vital information (e.g. on births, deaths) as they occur. These data files are clearly highly sensitive and are therefore not released but retained in a 'safe setting', typically within the national statistical office, with highly constrained access. Norway, Denmark, Finland, Sweden,

France and England and Wales all have these types of study – although population data from Denmark have not been obtained from questionnaires since 1970. Full details of internationally comparable data can be found in the *Longitudinal Study Newsletter* (Centre for Longitudinal Studies, 1999). Thus census microdata has developed along two parallel pathways, as 'safe data' which can be analyzed on the researcher's own PC and as data held in a 'safe setting' and often linked with administrative records (Marsh *et al.*, 1994). Data held in a safe setting, by virtue of the security which this gives, can retain much greater detail than data which have to be safe enough for wide scale distribution on the basis of a signed agreement.

Britain has experience of both these methods of providing security – the SARs and the ONS Longitudinal Study. Academic access to the latter is provided through a support unit funded by the Economic and Social Research Council at the Institute of Education (www.c/s.ioe.ac.uk). The support unit staff conduct analyses within ONS on behalf of academic users. By comparison with this 'safe setting' approach, many more people are able to obtain access to safe data, the cost of using the data is much less, and, in most cases, analysis is greatly simplified by the 'hands-on' approach. Whilst holding data in a safe setting allows much greater detail to be retained, the amount of support needed to run this kind of service can be very costly and may also impose additional restrictions in terms of the availability of software platforms and speed of turnaround (Marsh *et al.*, 1994). However, recent developments aimed at providing fast and effective delivery of output from safe settings may mean that this balance changes in the future (Keller-McNulty and Unger, 1998; Openshaw *et al.*, in press).

## 2.4  Methods of assessing the risk of disclosure posed by census microdata

All census offices go to considerable lengths to ensure that the chances of being able to identify an individual or household in a sample of microdata are negligible. But how is the risk of disclosure assessed?

Whilst different countries use different methods, there are some standard precautions that are widely adopted.

1. The sampling fraction is restricted. The higher the percentage of the population for whom records are released, the greater the risk of disclosure. There are several reasons for this. Firstly, by taking a relatively small sampling fraction, the chance that a given individual or household is in the sample is very small – perhaps 2 or 3 in 100. This not only reduces the number of individuals at risk, but also reduces the incentive for someone to attempt to identify a particular individual or to match the microdata records with another dataset.
2. The amount of detail available is restricted. In almost all samples of microdata there is a restriction on the level of geography released. It is generally agreed that knowing someone's area of residence adds significantly to the risk of identification and thus a threshold is usually implemented which, in the USA, Canada and Britain, is set at around 100 000 population. Detail on other variables may also be restricted by grouping or top-coding if there are small numbers in any categories. In Canada, the amount of occupational information released in the

PUMFs is limited to only 16 categories as this is seen to be another variable which facilitates identification. In Britain, much more detailed occupational information is available, but unlike in Canada there was, in 1991, no information on the individual or household income. However, income in the Canadian PUMFs was rounded so as not to exceed pre-set upper and lower limits.

Generally, these measures – sampling, restriction of geographical detail and grouping or rounding – are seen as providing sufficient protection. However, other measures involving perturbation of the data may also be used.

3. Perturbation measures include record-swapping, suppression and over-imputation. The 1991 Candian PUMFs used record-swapping, whereby a small number of records had their geographical identification changed to that of another area. This has been discontinued for 1996 files. In addition, for some records, data have been suppressed and a 'not available' code assigned.

The US 1990 PUMs used over-imputation for small areas where it was considered that the sample size did not provide adequate protection. A small subset of households was selected and some of the data items on the household records were blanked out. Responses to these items were then imputed using the same imputation procedures that were used for non-response (US Dept of Commerce, PUMS 1993, Ch. 3).

## 2.4.1 Formalizing disclosure risk

There is a considerable literature on methods of measuring and predicting the risk of disclosure. It is widely accepted that, before disclosure of information can occur, it is necessary for an individual in a sample of microdata to be identified. Marsh *et al.* (1991) argued that there are four factors which influence the probability of identification:

- the probability of being in the sample; this corresponds to the sampling fraction;
- the probability of key variables being recorded identically in the sample and the file being used to attempt identification;
- the probability of being unique in the population on the key variables;
- the probability of verifying population uniqueness.

In work to establish the risk involved in releasing the 1991 SARs for Britain, Marsh *et al.* (1991) made an assessment of these various elements and estimated that the probability of identification of a given record was negligible. This provided the assurances upon which the UK census offices accepted the case for samples of anonymized records.

Uniqueness is a key concept in statistical disclosure control for microdata. A person is *population unique* if he or she is the only person in a population with a certain combination of values for a set of *key variables*. Such a person is particularly at risk of identification if these variables are present in a released microdata file. The proportion of people who are unique in the population on a given set of keys has been widely used as a measure of disclosure risk (see, for example, Bethlehem *et al.*, 1990; Greenburg and Voshell, 1990; Müller *et al.*, 1992; Skinner *et al.*, 1994). In particular, Greenburg and Voshell (1990) have used it to assess the impact of geographic detail on risk and this work has been extended by Elliot *et al.* (1998) in the British context.

However, population uniqueness takes no account of the effect of the size of the sampling fraction on the risk of disclosure and thus the disclosure protection afforded by sampling. Therefore a second measure of risk may be used which combines information on both the sampling fraction and the level of population uniqueness: this is the proportion of sample uniques in a microdata file which are unique in the population. This measure has been considered in the context of the Canadian census by Carter *et al.*. (1991) and used by Dale and Elliot (1999) to produce an assessment of the relative risk attached to a range of specifications for samples of anonymized records from the 2001 Census.

Decisions over exactly how much detail can safely be included in samples of microdata are usually informed by investigations of the kind outlined above. However, it is also necessary to include an assessment of the protection afforded by various kinds of inaccuracies in the data (these include non-response, errors in reporting and recording, the effects of editing). Risk of identification is also reduced where the microdata sample and the external data source are not coded using the same classifications. (Even where classifications are the same, coding inconsistencies add further protection to complex variables such as occupation.) Additionally, households and individuals change their characteristics over time – for example, changes in occupation, work hours and place of residence, and, of course, there are changes in the composition of the household as children leave home, new children are born and partnerships are formed and dissolved. Finally, the availability of matching information in the country concerned and the motivation for making such an attempt influence the likelihood of making an attempt in the first place (Elliot and Dale, 1999).

All the evidence so far suggests that there has been no known attempt to identify individuals or households from records of microdata. However, it is a concern that is constantly present – increasingly so with the growing computational power of desktop computers – and reassessment of risk is required at each census.

## 2.5 The specific contribution of census microdata

In this section we identify those aspects of census microdata which make it uniquely distinctive from sample survey files and from aggregate census data for small areas. We illustrate these aspects by providing some examples of usage. For users new to microdata, it is, however, worth emphasizing the flexibility which it provides – for example, in being able to choose the hardware and software with which to analyze the data; in determining how to recode or regroup categories of variables; in the ability to relate variables from the whole range of questions asked on the census; and in using the range of multivariate methods available within statistical packages. In the following sections we review those features specific to samples of microdata drawn from the census of population.

### 2.5.1 Sample size

Census microdata files provide very large samples. The US 5 per cent PUMS contains over 5 million housing units and over 12 million individuals. Most countries have increased the size of their microdata samples over time and few release less than a 1 per cent sample. Almost without exception, this is larger than the biggest sample survey. In Britain, the Labour Force Survey has a sampling fraction of about 0.3 per cent and provides a hierarchical file with information on all individuals in the household. As such, it is less than one-third of the size of the 1 per cent Household File from the 1991 Census.

### The ability to analyze subgroups of the population

A large sample size is particularly important for the analysis of small population subgroups. Census microdata files often provide the only means of analyzing small groups with sufficient numbers to produce reliable estimates. In Britain, minority ethnic groups constitute only about 5 per cent of the total population. Because there are significant differences between ethnic groups in, for example, the level and reasons for unemployment, it is important to be able to distinguish between them. The British SARs have played an important role in identifying ethnic differentiation and highlighting incongruities between educational attainment and levels of unemployment (Blackburn *et al.*, 1997) and the occupational level achieved (Heath and McMahon, 1997). A full list of all publications based on the SARs is available on the website of the Census Microdata Unit (http://les.man.ac.uk/ccsr/cmu).

The data on ethnicity and income in the Canadian PUMF for individuals from the 1991 Census have been used to analyze the return to education in terms of income for different ethnic groups (Lian and Matthews, 1998). The 5 per cent sample of the US PUMS supports very specific analysis of small groups in the population. For example, Fang and Brown (1999) have used it to test theories of spatial mobility of foreign-born Chinese in three large metropolises – New York, Los Angeles and San Francisco. The files for 1970, 1980 and 1990 have been used by James (1998) to determine whether economic and demographic factors explaining entry into marriage among African-American men are constant over time. This research incorporated a multilevel analysis (see Ch. 7) to link macro-level indicators with individual level characteristics and marital outcomes. Bellente and Kogut (1998) used the 1 per cent PUMS to examine the effect of English language ability and time spent in the USA on the earnings of immigrants. Other work has used the PUMS to compare household versus individual approaches to understanding the migration of elderly people (Lin, 1997) and to analyze interstate retirement migration using both individual and locational characteristics (Clark *et al.*, 1996). The large size of the PUMS has allowed the analysis of single parent families headed by fathers and how they have changed between 1960 and 1990 (Eggebeen *et al.*, 1996). In Britain the availability of hierarchical data has allowed the first nationally representative study of family composition and partnership patterns amongst different ethnic groups (Holdsworth and Dale, 1995).

## 2.5.2 Level of detail

Microdata files provide detailed variable categorizations that allow analysts to make their own groupings or classifications. Whilst the amount of detail varies between countries, there is a marked contrast with aggregate census tabulations for small areas where all variables are precoded and little detail is available. Thus the 1991 British SARs contain 358 occupational categories on the 1 per cent file, which are grouped into 73 categories (SOC minor group) on the 2 per cent file. In the aggregate tables (SAS/LBS) there is only one table which provides the full 371 unit groups and it gives a cross-tabulation of employment status by sex only (SAS/LBS, Table 4). Even greater occupational detail is available in the US PUMS, as well as income, e.g. wages in dollars up to $140 000.

This sort of detail, together with the large sample size, provides the capacity for analysis of particular occupational groups or those who have a defined educational qualification (e.g. doctors or teachers). This can be important in manpower planning – for example, to provide an estimate of the extent of qualified labour in a country, where people with particular qualifications are located and what proportion are not currently in employment. Hakim (1998) uses the occupational detail of the SARs for Britain to identify pharmacists and to examine gender differences in their employment status.

In Britain, the detailed occupational and employment status variables in the Household SAR have been very important in providing the 'building blocks' which allow a range of alternative social classifications to be derived, based upon different ways of grouping these building blocks. When the SARs were released by ONS, they contained the social classification routinely used by government. However, other classifications were quickly added – Goldthorpe's social classification (Goldthorpe, 1987); the Women and Employment classification (Martin and Roberts, 1984); and the International Standard Classification of Occupations (International Labour Office, 1990) (see Section 2.5.6 for more details). This means that analysts can use a classification of most theoretical relevance to their research or they can choose a classification that facilitates international comparison.

## 2.5.3 Choice of unit of analysis and population

Microdata files allow analysts to choose their unit of analysis. In the USA, Australia and the UK, hierarchical files mean that there is a choice of working at the individual or household level and, in Australia and Britain, families can also be distinguished (Ch. 5 provides information on analyzing hierarchical data). The Canadian PUMFs provide three different files reflecting each of these units of analysis, but within each file there is additional information about the other two levels. This allows the analyst to work at whichever level is the most appropriate for the analysis.

Further choices arise over the population to be analyzed, e.g. whether a full age range is used or restricted groups. Thus children can be selected and analysis conducted of the circumstances of the families in which they are living. Alternatively, analyses can focus upon those aged 16–19 in order to examine who stays at school beyond compulsory school-leaving age (Drew et al., 1997). The British data allow analysis of inward migrants in the last year and have been used to add important evidence to the debate over the role of social (public) housing in restricting residential mobility and thus employment opportunities (Boyle, 1995).

Choices also arise over analysis of those living in private households or communal establishments. Census microdata files are unique in allowing analysis of those living in residential homes, hospitals, prisons or army quarters (discussed in the next section). Thus the analyst has a high degree of flexibility, although this also requires that decisions are made over selecting the most appropriate population and, if care is not taken, results may be very misleading. This ability to define the population base according to the research requirement is a unique aspect of census microdata and contrasts strongly with aggregate census tables where the unit of analysis and population are pre-specified and may not necessarily meet the needs of the analyst. These issues are discussed further in Chapter 5.

## 2.5.4 Coverage and sampling strategy

Related to the choice of population is the fact that the census covers the entire population – irrespective of geographical location, age, employment status or type of residence. Because samples of microdata are drawn from the census, they are based on a much more effective 'sampling frame' than is usually available to a social survey. In Britain, the Postal Address File (PAF) is routinely used as a sampling frame. Whilst it is estimated to contain 99 per cent of census addresses (Dodd, 1987), about 11–12 per cent of addresses are ineligible (Lynn and Lievesley, 1992). The Electoral Register, also widely used, is estimated to omit about 7 per cent of the eligible population and in Inner London this rises to 20 per cent. Differences by age and ethnic group are also very high – for example, 21 per cent of those aged 20–24 are omitted, as are 24 per cent of the Black population and 38 per cent in private furnished accommodation (Smith, 1993). The fact that the census is compulsory for the entire population means that all households should fall within the sampling frame.

Comparison may also be drawn between non-response in surveys and under-enumeration in a census. Checks on the extent and nature of under-enumeration were discussed in Chapter 1. Typical levels of under-enumeration of about 3 per cent are particularly troublesome when data are used at small areas for particular sub-groups. However, it is important to consider under-enumeration in a rather different way in the context of microdata where small areas cannot be identified. A comparison may be appropriate with the response rates of high-quality surveys where an 80 per cent response is considered acceptable. These, of course, are available only at higher geographical areas. For example, in Britain the General Household Survey has a sample size of about 12 000 responding households and a response rate of about 84 per cent which varies considerably for different groups of the population, with highest non-response amongst young people living in metropolitan areas, those who have recently moved, the self-employed and the unemployed.

The sampling design of most social surveys includes an element of stratification and a considerable amount of clustering – the latter in order to cut down costs. However, the method by which microdata samples are drawn from the census allows a high degree of stratification based on geographical proximity, with no requirement for any clustering. This is reflected in the size of design factors – the additional errors in a sample that are due to the design differing from a simple random sample. Therefore census microdata files should provide a better representation of the population and also have smaller design factors than most social survey data.

This means that census microdata are uniquely important for demographic research and population forecasting where a complete population sample is essential. The British SARs have been used in improving household projections (King and Bolsdon, 1998) in a number of ways – by testing sensitivity to the different ways in which lone parent households have been estimated; by allowing conventional projections to be disaggregated by age, gender, marital status and household composition; and by examining differences in the propensity to share a household and building that into projections for housing requirements.

The fact that census data include people living in institutions is of importance for a number of applications. Murphy and Wang (1996) use the SARs to make marital status population projections for England and Wales, where sample surveys cause problems because of the omission of the institutional population. Grundy et al.'s study (Section 8.2) of the health of the elderly provides a further example where the inclusion of the elderly living in residential care homes or hospital is essential for understanding marital status differences with age. Information on migration in the census can help in understanding differential movement of elderly people into residential homes.

The coverage of census microdata files, together with their geographical definition (discussed below), also makes them of great value in setting accurate quotas for many different kinds of surveys. Particularly in the world of market research, customers want surveys which relate to specific subgroups of the population – e.g. families with children under five; households with three or more cars – and these are often required to be specific to particular geographical areas, e.g. a large city or local district. Where small area data do not provide the required definitions and cross-tabulations, microdata files can be invaluable.

## 2.5.5 Geographical definition

The degree of geographical definition available is related to sample size and the reliability of estimates, as well as confidentiality. Generally, census microdata files provide much better geographical definition than sample surveys. The USA, Canada and Britain currently set minimum geographical areas of 100 000/120 000 population, which allows analysis at the level of large local government areas in Britain or by municipalities in Canada and counties in the USA. This is not, however, the case in Australia, where geographic information is limited to a State or Section of State classification in 1981 and 1986 files and to regions of 500 000 or more population in 1991.

Recent work in Britain suggests that the threshold of 120 000 in an individual-level file may be reduced to 60 000 without major implications for confidentiality (Dale and Elliot, 1999). The immediate research benefit from lowering the threshold would be the ability to identify more local government areas. Decisions over the geographic detail to be released in the UK SARs for 2001 may therefore reflect this research.

For local government, the availability of samples of microdata opens up a wide range of policy analysis. In Britain, local authorities (LAs) are used to using small area statistics (SAS) – a set of predefined tabulations for very small areas. However, the tables may not give exactly what is required; the variable groupings may be inappropriate or the combination of variables not what is needed. In these cases, microdata at the LA level can provide considerable extra flexibility. For example, they may be used to generate profiles of particular client groups within the authority and to support specific policy initiatives, e.g. care in the community (for the elderly or those with a long-term limiting illness). Microdata files also provide an opportunity for alternative ways of assessing deprivation, i.e. by identifying deprived households within areas, rather than deprived areas (Fieldhouse and Tye, 1996).

The size and unclustered sampling design of microdata files mean that geographical areas can be identified at a more detailed level than is the case with most sample surveys. This opens up the possibility of including 'place' in the analysis through multilevel modelling methods. Using the example of unemployment, area-level effects could be modelled most simply by assuming that overall levels of unemployment were greater for some areas than for others (for given individual characteristics). But the model can be developed further to allow different effects of the explanatory variables for different areas; and also cross-level effects which summarize the relationship between individual-level characteristics and an area-level characteristic (Jones, 1997; Fieldhouse and Gould, 1998).

If the geographical areas available in the census microdata file are appropriate to the analysis, then including both individual and area-level information can be a very powerful method. However, for Britain, the most detailed geography available in the SARs does not go below local government districts and these do not always correspond to local labour markets. Fieldhouse and Gould (1998) provide an example of multilevel models of unemployment where aggregations of local government areas are used as approximations to local labour markets.

An innovative method of extending the value of geographical definition whilst retaining confidentiality has been adopted for the British SARs. This has involved the addition of an area-level classification to each individual (Dale and Openshaw, 1997; http://les.man.ac.uk/ccsr). This provides additional geographical information without revealing the geographic location. For example, the Individual SAR for 1991 identifies the city of Manchester but does not allow areas within the city to be identified. However, each individual in the Manchester SAR has been allocated to one of 49 different categories of an area-level classification, GB Profiles (Openshaw *et al.*, 1994). Thus residents of Manchester may be analyzed in terms of the characteristics of their locality. The value of this in a multilevel modelling framework is that it allows type of locality to be assigned within the identified geographical area.

## 2.5.6 Addition of derived variables

This detailed individual-level information allows new classifications to be derived by the analyst – for example, a range of different social classifications have been added to the UK SARs based on combinations of occupation and employment status (Section 2.5.2). The same information allows the addition of occupational status scores, derived from other studies, but also based on occupation and employment status. These include the Cambridge Occupational Scale Score (Prandy, 1990, 1992), a continuous measure, designed to measure social advantage and based on the assumption that there is social interaction between those with similar lifestyles. Other measures added are the Standard International Occupational Scale (SIOPS) (Ganzeboom and Treiman, 1995), based on prestige evaluations carried out in approximately 60 countries, and the International Socio-Economic Index of Occupational Status (ISEI) which measures the attributes of occupation that convert a person's education into income (Ganzeboom *et al.*, 1992).

A similar example comes from the addition of earnings information to the SARs. Mean hourly earnings were derived from an employer-based survey of incomes in the form of a table broken down by variables such as age, sex, full- or part-time working and region. The availability of all these variables on the SARs, coded in the same way in both studies, allows the addition of this 'earnings score' to all individuals in the SARs who report an occupation. The addition of an area-level classification to individuals in the British 1991 SARs was described in the previous section.

Where microdata files are organized hierarchically (US, UK, Australia) there is additional scope for deriving many new variables which summarize the characteristics of the household or family. For example, household classifications can be derived which take into account information for all members of the household. These can be fine-tuned to reflect the particular focus of the research – for example, a classification designed for a study of housing conditions might make key distinctions between single-person households and couples with and without children. Identifying one-parent households might also be important. A rather different classification might be derived for a study of the household composition of elderly people, with categories devised to reflect the household distinctions of most importance for the elderly. The range of variables in the census allows indices of housing needs or of 'material assets' to be developed and related to household composition (Dale *et al.*, 1996). Section 5.2 provides details of how to derive these variables.

## 2.6 The relationship between microdata and small area data

Many census users are very familiar with small area data such as the SAS for the GB census, but less so with microdata. At heart, small area data are appropriate where the unit of analysis is the area (at whatever spatial level is decided), whilst individual-level data are preferable where the analysis relates primarily to individuals and place is of secondary concern.

This can be simply demonstrated with reference to an analysis of unemployment. Unemployment shows considerable variation, with very high pockets of unemployment often concentrated in very specific areas, e.g. on housing estates. If the requirement is to identify the areas of highest unemployment in order to target resources (e.g. drop-in centres) then small area data are essential. The most appropriate measure for identifying need is the subject of continuing debate. Simpson (1996) discusses the adequacy of a range of standard methods, including the signed chi-squared, and shows the properties of each measure.

However, if the requirement is to identify the determinants of unemployment, making an assumption that individual-level characteristics (e.g. age, educational qualifications, ethnic group) will be of primary importance, then the application of even straightforward multivariate analysis is hampered unless microdata are available. For example, unemployment varies with age and gender and there is, typically, considerable difference between ethnic groups in their demographic profile. Educational level is known to be an important predictor of unemployment, and family responsibilities may also be important. Thus an adequate model would include at least five variables – employment status, age, sex, educational level and ethnic group – with additional information on family responsibilities desirable and, perhaps, whether or not a first-generation immigrant.

Small area data do not provide this level of detail in a single table and to attempt this analysis using aggregate data drawn from two or more tables will invariably raise problems of the ecological fallacy. To establish the degree of association between two variables – e.g. unemployment and the level of educational qualifications – if these variables are not available in the same table, necessitates the use of correlations based on aggregations at the smallest possible level. Seminal research by Robinson (1950) using the association between literacy rates and race in the USA demonstrated that this approach can produce seriously misleading results. In this case, there was a strong positive correlation at the level of large administrative areas between literacy rates and the percentage of the population that was black. However, when the analysis was repeated at the individual level this association fell to 0.02. The explanation lay in the fact that black people tended to live in the same areas as poorly educated whites. It was the latter group which was most likely to have children with low levels of literacy. However, at the aggregate level this could not be established.

## 2.7 Looking backwards

A topic of considerable importance is the issue of preservation of census materials and access to historical data. Many of the problems of confidentiality become less important as data get older and this increases the scope for microdata samples. In the USA researchers at the University of Minnesota have constructed 25 public use microdata samples which span the censuses of 1850 to 1990 (www.ipums.umn.edu/), whilst in Canada, a project has been conducted for a part of the maritime region using historical documents from the National Archives of Canada.

The UK has a 100-year rule which means that 100 years after their collection census records may be released. This has led to the creation of a 2 per cent sample of records from the 1851 Census and a computerized transcription of the census enumerators' books for the 1881 Census for England and Wales, the Channel Islands and the Isle of Man. The latter dataset consists of the name, address, relationship to the head of family, marital status, age, occupation and birthplace of some 26 million individuals, together with information about disabilities and any other notes recorded in the census. It is a by-product of a project to create a microfiche index of the population of Great Britain by the Genealogical Society of Utah. More details can be obtained from The Data Archive (http://dawww.essex.ac.uk).

## Appendix 2.1: Details of census microdata files for the USA, Canada, Australia and the UK

## Public use microdata files in the USA

The US Bureau of the Census released three different files from the 1990 Census:

- A 5 per cent sample identifying every state and subdivisions of states with at least 100 000 persons. This means the public use microdata areas (PUMAs) are primarily counties, but in some cases may be groups of counties or places. Nationwide this gives a sample of over 12 million persons and over 5 million housing units.
- A 1 per cent sample identifying metropolitan areas and non-metropolitan areas (or a mixture of the two). Metropolitan authorities (MAs) with 100 000 or more inhabitants are identified individually, but are otherwise grouped.
- A 3 per cent elderly sample, with the same output geography as the 5 per cent sample.

The US PUMS contain the full range of population and housing information collected in the 1990 Census. This covers about 25 housing/householder topics and about 40 person topics, although the actual number of variables is considerably greater. The PUMS retain a high degree of variable detail, including, for example, 500 occupational categories, age by single years up to 90, and wages in dollars up to $140 000. The files also contain many of the more commonly used derived variables, e.g. household income and poverty status. Individual weights are also available to allow the user to approximate the distributions in the published tables.

The files are structured with a record for each housing unit, followed by a variable number of records for each person in the housing unit. A housing record serial number of both the housing record and the person record allows linkage of individuals within the housing unit (see Fig. 2.1).

### Sampling design

The sample for the 1990 PUMS was drawn from households that responded to the long form in the census, and based on housing units, except for institutions, where persons were sampled. A stratified systematic selection procedure with equal probability was used to select each of the public use microdata samples. The sampling universe was defined as all occupied housing units including all occupants, vacant housing units and group quarters (GQ) persons in the census sample. The sample units were stratified during the selection procedure in order to improve the reliability of the estimates of the PUMS by increasing within-state homogeneity with respect to characteristics of major interest.

Units were divided into households, vacant housing units and GQ population. The household universe was stratified by family type and non-family, race/Hispanic origin of the householder, tenure, and age within sampling stratum. The vacant housing units universe was stratified by vacancy status and sampling rate. The GQ population

## SUMMARY DATA

- Basic unit is an identified geographic entity
- Data summarized on people and housing in specific entity
- Available for small areas

**ILLUSTRATIVE SUMMARY DATA**

| Place | Total population | Occupied housing units | Persons per unit | Renter occupied units | Gross rent Under $100 | $100–149 | $150–199 |
|---|---|---|---|---|---|---|---|
| Weston City | 110 938 | 49 426 | 2.2 | 31 447 | 158 | 1967 | 6282 |
| Smithville | 21 970 | 7261 | 3.1 | 2492 | 17 | 90 | 766 |
| Junction | 17 152 | 5494 | 2.7 | 822 | 11 | 29 | 238 |

## PUBLIC USE MICRODATA

- Basic unit is an unidentified housing unit and its occupants
- Unaggregated data to be summarized by the user
- Allows detailed study of relationships among characteristics
- Not available for small areas

**ILLUSTRATIVE MICRODATA***

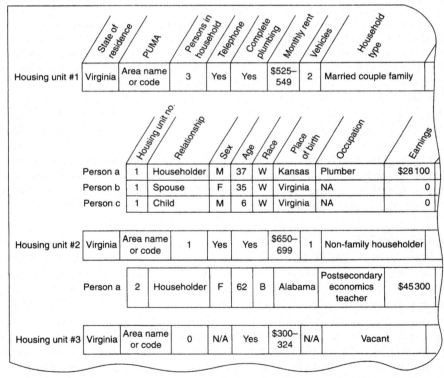

* Public use microdata samples do not actually contain alphabetic information. Such information is converted to numeric codes; for example, the State of Virginia has a numeric code of 51.

**Fig. 2.1** Comparison of summary data with information on microdata files. Reproduced from the technical documentation on the US PUMS Figure 5-10, US Census of Population and Housing, Guide, Part A. Text, Bureau of the Census

was stratified by GQ type (institutions, non-institutions), race, Hispanic origin and age.

The sampling for the PUMS was then done as follows. The 1 per cent PUMS were drawn on a state by state basis. The sampling factor used depended upon the percentage in the state which completed the long form. If the long-form sample was 20 per cent, then this 20 per cent was divided into 20 1 per cent samples of equal size and one was selected at random.

Of the remaining 19 subsamples, five were selected at random and merged to form the 5 per cent PUMS. The 3 per cent elderly sample was produced in the same way but required an extra step. The three subsamples were merged and then households with at least one person aged 60 or more (or GQ persons aged 60 or more) were selected and designated as the elderly PUMS file.

## Confidentiality constraints

The confidentiality of the US PUMS is governed by Title 13 of the United States Code. For most areas it was assumed that the sample size, together with a limit on the geographic detail, provided sufficient protection to ensure that information was not disclosed about specific individuals, households or housing units. For a geographical area to be fully identified, there had to be a minimum population of 100 000.

However, for small areas where it was considered that the sample size did not provide adequate protection, a confidentiality edit was applied. A small subset of households were selected and some of the data items on the household records were blanked out. Responses to these items were then imputed using the same imputation procedures that were used for non-response (US Dept of Commerce, PUMS 1993, Ch. 3).

Using this method, the data structure is preserved whilst some additional protection is added. Finally, certain variables are top-coded or their actual values replaced by descriptive statistics such as the median.

## Cost and availability

Costs of the 1990 data and the basis on which they are available are given as indicative only. Both may be very different for the 2000 Census.

Distribution of the US PUMS is handled by the US Bureau of the Census with cost according to file size. For the 1990 PUMS, the complete 5 per cent file costs $4000 and the 1 per cent file costs $800. The New England division of the 5 per cent file costs $275, and the 1 per cent file costs $175.

The files can be purchased without any licensing procedure and are available to any organization or private individual. There is therefore no restriction on use outside the USA. The data files are also deposited at the Inter-University Consortium for Political and Social Research (ICPSR) at the University of Michigan where they can be obtained free of charge for academic use. UK users are able to order the PUMS through The Data Archive at the University of Essex (http://dawww.essex.ac.uk).

Costs of the PUMS from the 2000 Census, and the basis on which they are made available, may be different. Under current plans, 2000 PUMS should be available via the internet, using a new data retrieval system called the *American FactFinder* (AFF), by which the US Bureau of the Census plans to release much of the 2000 Census output. Latest information on this and other developments for Census 2000 are available from the following contacts:

> US Bureau of the Census
> Web page: http://www.census.gov
> E-mail: 2000usa@census.gov
> Microdata contact: Amanda Shields: 301457 1326

# Public use microdata files in Canada (PUMFs)

Statistics Canada have produced PUMFs for every quinquennial census since 1971, retaining a similar structure throughout the time period (Albert *et al.*, 1996). The PUMFs consist of three files structured around three units of analysis: the individual, the family and the household. Each file is rectangular, but each contains some summary contextual information about the other two units of analysis. A 1 per cent sample was released for each file from the 1971 Census; this increased to 2 per cent for the individual file in the 1980s and to 3 per cent for all files for the 1991 and 1996 Censuses.

There were six PUMFs for the 1971 Census: three separate files from a national sample for individuals, families and households; and then three other files for census metropolitan areas (CMAs) at the individual, family and household levels. Starting in 1976, Statistics Canada integrated the CMA sample within the individual, family and household files by including a variable to identify CMAs.

The following information is based on documentation for the PUMFs from the 1991 Census. However, where available, this is updated to reflect the latest web-based information on the recently released files for 1996.

## Population coverage

The population coverage is all Canadian citizens and landed immigrants with a usual place of residence in Canada or those who are residing overseas on a military or diplomatic placement. The 1991 file also contains non-permanent residents in Canada. Residents of institutions, foreign residents and residents of incompletely enumerated Indian reserves are not included.

## Structure and content of 1991 PUMFs

- The 3 per cent individual sample. This file contains over 119 variables for 809 654 individuals and includes sociodemographic information, place of birth, immigration, ethnic origin, language, educational qualifications, employment, occupation, industry, income, some housing information and a few summary variables about the household and family characteristics. Provinces are identified and CMAs with a population of at least 100 000 at the previous census. (Of the 25 CMAs in Canada, only 19 were categorized in the 1991 PUMF CMA variable because of grouping. For example, the CMAs of Regina, Saskatchewan, and Saskatoon,

Saskatchewan were combined into one CMA. These two cities are 140 km apart, separated by the open prairie.)

- The 3 per cent family file – containing 345 351 records and 138 variables. The family file contains demographic, social, cultural and economic information for families, their members and for non-family persons. Records for both partners in a family are present on many individual-level variables, e.g. occupation, age and fertility. Similar information is held for non-family persons. Information on housing is also available and geographic information for provinces, selected CMAs and municipalities.
- The 3 per cent household file – containing 298 960 records and 124 variables. The file contains summary information on the household – for example the total income, number of earners, household composition; socioeconomic and demographic information on the 'household maintainer' and his/her spouse; information about the maintainer's economic family; and housing information. Geographical information is limited to the province, and urban/rural code and CMAs. The population covered is all private households.

Occupational information is much more limited than for the US PUMS or the British SARs (only 16 occupational categories), although the files also contain detailed income information from a range of different sources.

## Sampling design

The PUMFs were sampled from the 20 per cent of the population which answered the long questionnaire. This sample was divided into three frames: one for each of the samples – household, family and individual. Within these frames, records were stratified geographically, and by further variables relevant to each of the three files.

## Confidentiality constraints

Statistics Canada have set up a Microdata Release Committee which reviews all proposals to release microdata files. Release of data is governed by the 1971 Statistics Act which prohibits the publication of any personal information from which an individual could be identified.

In order to preserve confidentiality, restrictions are imposed on the level of geographical details released, and some variables are grouped and top-coded. In the 1991 PUMFs, a certain amount of record-swapping took place although this has been discontinued for 1996. For some records, data have been suppressed and a 'not available' code assigned and income has been rounded not to exceed preset upper and lower limits.

## Costs and availability

For any first file the cost is $1000.00; for the second file it is $300.00 and for the third file $200.00.

A microdata licence agreement is required before the purchase and delivery of this product:

Household PUMF 1991:   Catalogue No. 95M0008XCB   1996
Family PUMF 1991:      Catalogue No. 95M0009XCB   1996: No. 95M0014XCB
Individual PUMF 1991:  Catalogue No. 95M0007XCB   1996: No. 95M0010XCB

The descriptions for the above products can be viewed on the web at the following URL – http://www.statcan.ca/english/search/ips.htm – which is the catalogue for information on products and services. Enter the catalogue number above for any of the three products to get more information about them. The record layout for each of the three files is available through the thematic search tool at the URL http://www.statcan.ca/english/search/.

## Academic access

The Canadian experience provides an interesting example of how academic access to census data can be opened up. In 1984 the newly elected federal government cancelled the 1986 Census as part of a cost-cutting exercise (Albert *et al.*, 1996). When it was reinstated it was on the basis that many of the products from the Census, including the PUMFs, would be produced on a cost recovery basis. The cost of the Canadian microdata files was much greater than the marginal costs charges in the USA for the PUMS and this resulted in many Canadian academics using US rather than Canadian data. However, the Canadian Association of Public Data Users (CAPDU), the Canadian Association of Research Libraries (CARL) and five federal departments formed a consortium and shared the costs of producing and disseminating the 1986 PUMFs (Humphrey, 1991; Albert *et al.*, 1996) and a similar consortium purchase was used for the 1991 PUMFs. In January 1996, a new programme was introduced for post-secondary institutions within Canada, providing access to Statistics Canada 'standard data products' on a subscription basis. Known as the *Data Liberation Initiative/Initiative de Démocratization des Données* (DLI/IDD), this programme offers full access to all 1996 Census data products, including the PUMFs. The earlier experiences with the CARL data consortium helped pave the way for DLI/IDD. First, the CARL consortium was a library initiative and legitimized digital data in the eyes of library directors (some directors had the preconceived notion that only books belonged in libraries). Once data became an acceptable library product, CARL directors were willing to seek better access to Statistics Canada data products. Secondly, Statistics Canada also learned from the CARL consortium of the early 1990s. They observed that libraries were capable of housing their data products and offering client support services in the use of these data. Thus, the consortium served to educate both senior management in the library community and Statistics Canada that access to data was a positive venture in which to be engaged.

# Census microdata files in Australia

The Australian Bureau of Statistics (ABS) produced a Households Sample File (HSF) and a Persons Sample File from the 1981 Census, two hierarchical HSFs from the 1986 Census, and a single hierarchical HSF from the 1991 and 1996 Censuses. There are plans to release a 2001 HSF by 2003. All HSFs include records for all persons in a 1 per cent sample of households and a 1 per cent sample of persons in non-private dwellings. Hierarchical files include dwelling, family and person records. The 1981 files have a flat file structure but include dwelling and family variables at the person level, with household and family identifiers on the HSF allowing persons within families and households to be related.

### Structure and content of the 1991 HSF

The 1991 HSF contains information collected on census night, 6 August 1991, from a 1 per cent sample of households in occupied private dwellings and all persons counted within each selected dwelling, and a 1 per cent sample of persons in non-private dwellings. The file is hierachical, with dwelling, family and person record types. Within a dwelling (or household), there may be up to four family records and their associated persons records, corresponding to the primary, second and third family and any non-family adult residents.

There are 58 524 households containing 62 503 families and 162 736 family and non-family members, and 5805 residents of non-private dwellings. The full range of census variables is available, although with some reduction of classificatory detail for some variables and limited geographic information.

### Sampling design

The sample of private dwellings is a systematic selection of all occupied private dwellings identified in the Census, ordered geographically to ensure appropriate representation by state and territory and by region. Similarly, the 1 per cent sample of persons in non-private dwellings is a systematic selection of individuals from all persons identified within such dwellings.

### Confidentiality constraints

Sample files of census unit record data are released in accordance with a Ministerial Determination under Section 13 of the Census and Statistics Act, which requires users to sign an undertaking stating that the information will be used for statistical purposes only. In addition, the ABS established an internal Microdata Review Panel which seeks to ensure that 'information is not likely to enable the identification of the particular person or organization to which it relates', as required by the Determination.

Geographical information in this file differs from previous census releases, being based on regions with a population of 500 000 or more. This gives more detail in the major states of New South Wales, Victoria and Queensland, with seven, five and four categories, respectively, and identifies the capital cities in the five mainland states. However, the remaining non-metropolitan populations of South Australia, Western Australia and the Northern Territory are combined in one category, with Tasmania and the ACT in the final grouping.

Categories of occupation, industry, highest qualification, country of birth of individual, mother and father, and age have been grouped for confidentiality but retain the basic structure of the classifications and a reasonable level of detail. Grouping in other variables such as income, rent and mortgage payments resulted from the change to an optimal mark recognition (OMR) design for the census form.

### Costs and availability

The 1991 HSF is available to Australian educational institutions through the Social Science Data Archives (SSDA) at the Australian National University at the reduced price of $2000. Copies of the file obtained under this arrangement can be used for publicly funded teaching and research only, and users of the file under these arrangements are required to sign an undertaking to this effect.

Other organizations may purchase a copy of the 1991 HSF from the ABS for $5500 and are required to sign an undertaking stating that the data will be used for statistical purposes only.

A copy of the *1991 HSF User's Guide* is available on-line and can be downloaded in rich text format (RTF) from the SSDA's website on http://ssda.anu.edu.au. To contact the SSDA, phone + 61 6 249 4400, fax + 61 6 249 4722 or e-mail ssda@anu.edu.au.

For the ABS Census Marketing Unit, ring (freecall) on 1800 813 939, fax + 61 6 251 3352 or visit the ABS website on http://www.abs.gov.au.

### 1996 HSF

The structure of the HSF for 1996 is basically the same as that for 1991, with small modifications reflecting user's preferences and ABS confidentiality requirements. At the time of writing, the 1996 HSF is not deposited in the SSDA. For latest information on file details and availability, consult the ABS web pages. The cost of the file is $7500.

# Samples of Anonymised Records in the UK

In Britain, samples of microdata were produced for the first time following the 1991 Census and here, too, there is a hierarchical file. Requests had been made for SARs to be released from previous censuses in Great Britain. The principal stumbling block in the past had been an argument as to whether SARs could be considered a statistical abstract for release under Section 4.2 of the Census Act 1920 at the request and expense of user(s).

## Structure and content of the SARs

Two SARs have been extracted from the censuses for England and Wales, Scotland and Northern Ireland:

- *The Individual SAR* – a 2 per cent sample of individuals in households and communal establishments; this comprises 1.1 million records containing information on:

    - individual characteristics, e.g. age, sex, employment status, occupation and social class;
    - accommodation, e.g. availability of a bath/shower and the tenure of the accommodation;
    - limited information about the individual's family head – sex, economic position and social class;
    - limited information about other members of the individual's household (such as the number of persons with long-term illness and numbers of pensioners).

The finest geography is the local authority, with a minimum population size of 120 000. Subthreshold areas are grouped with neighbouring areas.

- *The Household SAR* – a 1 per cent hierarchical sample of households and individuals in those households; this comprises 215 000 households and the 542 000 individuals enumerated in them. The full range of census variables is available, with standard region as the lowest level of geography.

Similar files are planned for the 2001 Census, although there are proposals to increase the sample size of the Individual SAR to 3 per cent and to reduce the population threshold.

## The sampling procedure used

In 1991 the Census went through two separate coding processes. The easy-to-code information, such as housing details, sex, date of birth, and country of birth, was processed for all forms (100 per cent). The harder-to-code information such as occupation and industry was only processed for 10 per cent of forms. Both SARs were drawn from the 10 per cent sample so that they contain information from the whole of the census form. A detailed description of the sampling scheme for the SARs is given in Dale and Marsh (1993, Ch. 11).

## Confidentiality protection in the SARs

In the 1991 SARs, an area had to have a population size of at least 120 000 in the mid-1989 estimates to be identified separately. The primary units used were local districts; only one geographical scheme was permitted, or smaller areas could be identified in the overlap, say between a local district and a health district. A population size of 120 000 is slightly higher than the lowest level of geography permitted in the US SARs (100 000), but it still has the advantage of allowing all non-metropolitan counties in England and Wales, most Scottish regions, all London boroughs (except the City of London), and all metropolitan districts to be separately identified. Smaller local authority districts (under 120 000 population) were grouped to form areas over 120 000.

The 1 per cent Household SAR, because of its hierarchical nature (i.e. statistics about the household and all its members), is more of a disclosure risk. For this reason it was decided that, for this SAR, the lowest geographical detail revealed would be the Registrar General's Standard Regions, plus Wales and Scotland. The only exception is that the South-East is split into Inner London, Outer London, and the Rest of the South East Region.

The order of records in both SARs was rearranged before the Census Offices released them, to prevent any possibility of tracing individuals or households back through a region or district.

When expected frequency counts fell below a predefined threshold, categories were grouped. With some variables, grouping was only required at one end of the distribution: thus rooms were top-coded above 14 and the number of persons in the household was top-coded above 12. Two variables were both grouped and top-coded; with age, 91 and 92 were grouped, 93 and 94 were grouped and 95 and over was top-coded; with hours of work, 71–80 hours per week was grouped and the rest top-coded above 81.

There were other factors which determined the detail to be released:

- Categories of occupations and industries in the public eye were grouped further than mathematically necessary to guard against disclosure.
- Large households were seen as a disclosure risk in the household sample. Consequently, only housing information is given for households containing 12 or more persons. No information about the individuals in the household is given.
- Geographical information for such items as workplace and migration (address 1 year before census) has been heavily grouped. This is because of the high likelihood of uniqueness of such information when used in conjunction with area of residence.

## Costs and availability

The 1991 SARs are available for academic use within the UK free of charge. For academic use within the UK there is a two-stage registration procedure: Higher education institutions (HEIs) are required to sign an 'end user licence agreement' which makes the HEI responsible for those members of their institution who are using the data. Then, users within each institution – who must be either members of staff or students – sign a further individual registration form which contains a binding undertaking to respect the confidentiality of the data. Specifically, users have to undertake not to attempt to obtain or derive information about an identified individual or household, nor to claim to have obtained such information from the SARs. Furthermore, they have to undertake not to pass on copies of the raw data to unregistered users. Overseas academics can apply to use the 1991 SARs on a similar basis to academics in the UK.

Non-academic organizations may purchase copies of the SARs and sign a similar end user licence agreement and undertake not to allow the data to be used other than by their employees.

Similar arrangements are envisaged for 2001 SARs, except that UK academics should have a simplified registration procedure with a single registration allowing use of all 2001 Census products.

The licensing and distribution of the 1991 SARs are the responsibility of Manchester University who have a contract with the ESRC. The SARs are unique in having a unit with the specific responsibility of support and distribution. Charges for non-academics wishing to buy the data files are £1000 + VAT for the entire national SAR, and £500 for subsets of a county or local district.

For GB and Northern Ireland, contact:

> The Census Microdata Unit
> Manchester University
> Manchester M13 9PL
> Phone: +44 (0)161 275 4721
> Fax: +44 (0)161 275 4722
> E-mail: ccsr@man.ac.uk
> Web page: http//les.man.ac.uk/ccsr

**Table A2.1**  Websites for European national statistical institutes

| | |
|---|---|
| Austria | http://www.oestat.gv.at/index.htm |
| Belgium | http://www.statbel.fgov.be/ |
| Bulgaria | http://www.acad.bg/BulRTD/nsi/index.htm |
| Croatia | http://www.dzs.hr/ |
| Cyprus | http://www.pio.gov.cy/dsr/ |
| Czech Republic | http://www.czso.cz/ |
| Estonia | http://www.stat.ee/ |
| Hungary | http://www.ksh.hu/eng/nepszaml.html |
| Denmark | http://www2.dst.dk/internet/startuk.htm |
| Finland | http://www.stat.fi/ |
| France | http://www.insee.fr/ |
| Germany | http://www.statistik-bund.de/ |
| Greece | http://www.statistics.gr/ |
| Hungary | http://www.ksh.hu/ |
| Iceland | http://www.statice.is/ |
| Ireland | http://www.cso.ie/ |
| Italy | http://petra.istat.it/ |
| Latvia | http://www.csb.lv/ |
| Lithuania | http://www.std.lt/ |
| Luxembourg | http://statec.gouvernement.lu/ |
| Netherlands | http://www.cbs.nl/ |
| Norway | http://www.ssb.no/ |
| Portugal | http://infoline.ine.pt/ |
| Poland | http://www.stsp.gov.pl/ |
| Romania | http://cns.kappa.ro/ |
| Russia | http://feast.fe.msk.ru/ |
| Scotland | http://www.open.gov.uk/gros/groshome.htm |
| Spain | http://www.ine.es/ |
| Sweden | http://www.scb.se/ |
| Switzerland | http://www.admin.ch/bfs/ |
| Turkey | http://www.die.gov.tr/ |
| Yugoslavia | http://www.szs.sv.gov.yu/homee.htm |

# 3

# Cross-national comparisons

## 3.1 Introduction

In Chapters 1 and 2 we discussed the value of census data in general and microdata files in particular. In Chapter 2 we reviewed the availability of microdata files across Europe and North America. Although many countries do not allow census microdata to be sent overseas for use by other researchers, there are a growing number of countries where this is permitted, including the USA, Canada, Australia and Spain. This opens up the opportunity for cross-national comparative research and, even where microdata files are not allowed to go outside the country, comparative research can be profitably pursued by collaboration between researchers in different countries. In fact the research design will almost certainly be enhanced by such collaboration. Therefore this chapter is concerned with highlighting the research potential of cross-national comparisons, discussing some of the issues that need to be confronted before the analysis stage, and offering guidance in resolving difficulties which arise during analysis. We begin with a brief review of the various research designs typically used in comparative analysis and then move on to focus specifically on comparative analysis using census microdata.

## 3.2 Why make cross-national comparisons?

There is a growing interest in examining the differences between countries and understanding the reasons for them. From a basic descriptive standpoint, the growth of supranational bodies such as the United Nations and the European Union has led to a wealth of comparative statistics, much based on census data. This data plays an important role in identifying differences in the industrial or economic base of countries, and in providing demographic and labour market comparisons – see, for example, EUROSTAT's *Europe in Figures* and *Statistics in Focus* (http://europa.eu.int/en/comm/eurostat/), the *UN Statistical Yearbook* published annually (http://www.un.org/), *The World Bank Atlas* (1998, 30th edn) (http://worldbank.org/) and the World Health Organization (http://www.who.org/). Whilst a great deal of effort goes into ensuring a comparable base for cross-national statistics, there are always major difficulties in making face-value comparisons between published aggregate statistics for different countries. Nonetheless, these published statistics are heavily used in the policy arena to make comparisons in terms of, for example, infant mortality, disposable income, educational levels, fertility, age structures and life expectancy. The position of a country with respect to one or more of these measures may be used as a basis for policy formulation nationally and also at a supranational level. However, the availability of microdata opens up the possibility of much more detailed comparative analysis that moves beyond the superficial presentation of published tables to look more carefully at the underlying processes.

Most social research is comparative in one way or other. For example, we make comparisons between men and women, between different social classes or ethnic groups and between geographical areas within the same country. There is, therefore, some difficulty defining the limits and essential characteristics of comparative research. At a very basic level, however, it may be defined as based on data from at least two societies, and, in making international comparisons, these groups are usually defined as nation states.

In the following section, we review the various reasons for making cross-national comparisons and the research design which this entails.

## 3.2.1 Theory development and theory generalization

One of the important roles of comparative analysis is in either developing theory or generalizing theory. The assumption that lies behind this is that, if a theory is to have lasting value, it will identify relationships that are sufficiently fundamental to hold across countries at a similar level of industrialization, or with similar historical or cultural contexts.

Theory may be tested by establishing whether predicted relationships are found across a number of comparable countries – whether there is a 'universal' truth which holds despite national differences in culture or social policies (Scheuch, 1990). For example, Erikson and Goldthorpe (1992) used microdata to test the hypothesis advanced by Featherman et al. (1975) that the overall pattern of intergenerational mobility is similar across industrialized societies. They used data from 11 nations to test this hypothesis of 'common social fluidity'. Access to micro-data was crucial because it allowed the development of a comparable social classification based on detailed occupational classifications for individual countries and also enabled calculation of relative (net of the effect of structural features) rather than absolute mobility rates. Whilst generally Erikson and Goldthorpe's results supported the idea of a core pattern of social fluidity, they found that there was variation around this core. For example, Sweden appeared to have greater fluidity that was increasing over time.

In economics, universal laws may be expected to persist across national boundaries. Oswald (1997) tested the relationship between satisfaction or 'happiness' and economic performance using the Eurobarometer Survey series for 1973 onwards. Whilst the unemployed are very unhappy, he concludes that the happiness of people living in industrialized countries is not completely associated with improvement in economic performance. Despite dramatic improvement in quality of life in recent decades, levels of life satisfaction have increased only marginally.

Another example where comparison between countries is used to identify common elements that transcend the national level is given by Szakoiczai and Fustos (1998) and is based on data from the World Values Survey. An analysis of value preferences, using individual-level data for 24 countries in western and eastern Europe, tests the hypothesis that social background factors influence value preferences not through economic variables or through the extent of liberalization under communism, but through 'the stamps of axial moments like Protestantism, the Enlightenment and the different versions of socialism' (Szakoiczai and Fustos, 1998).

Comparison between countries may be used to refine and test further a hypothesis. Thus, the differences between two or more countries may be used to better understand the process of interest. For example, Jonsson and Mills (1993) used the work of Erikson and Goldthorpe to develop a more detailed set of hypotheses for Sweden and England in order to identify the factors that promote or impede social fluidity. Although they urge some caution in their results, they suggest that the higher level of fluidity in Sweden supports the idea that strong working-class organization and long-standing egalitarian policies are likely to promote equality of opportunity.

Alternatively, cross-national comparisons may be used to establish the generality of a theory that has been developed within a national context. This theory would be developed and refined by moving from a national to an international level. For example, in the UK we find that children from middle class backgrounds are more likely to go to university than are children from the working classes and this difference is not decreasing with the expansion in higher education. To test the generality of this finding, one might want to identify one or more countries with different ways of providing access to higher education as critical cases. For example, Egerton (1993) selected the USA and Hungary – the former classified as free market and the latter as a centrally planned system – and sought to explore differences in the effect of parental cultural capital on the educational attainment of children.

In these two examples, the countries being compared were carefully chosen to maximize differences in the characteristics that are hypothesized to influence the outcome. These differences may either provide a better understanding of the conditions under which a relationship holds or test whether a relationship observed in one national context is still found in a different one.

## 3.2.2 Understanding national differences

A major reason for international comparative research is to identify specific attributes that may explain observed differences in outcomes between two or more countries. For example, a great deal of comparative research has focused on differences in women's employment patterns in western industrialized countries (Glover, 1992; Rubery et al., 1998). The aim of these analyses is usually to explain the observed differences, using a theoretical framework that can be applied across different nation states. Thus Rubery et al. (1998) compare European trends in work patterns and working time. By focusing on specific relationships – e.g. national differences in working hours of mothers – they are able to classify countries which exhibit similar patterns and then seek explanations by examining differences in household and institutional arrangements. Pffau Effinger (1998) provides a theoretical framework to explain differences in women's part-time employment in the Netherlands, Germany and Finland which includes three basic concepts: gender order, gender culture and gender arrangement. These allow distinctions between cultural and institutional factors and also include the role of collective and individual social actors. The application of such a theoretical framework recognizes the complexity of influences on part-time employment – most of which cannot be formally tested in an empirical model using microdata. These examples are not designed to extend or test theory, but to apply a theoretical framework to understand observed differences between countries.

Holdsworth (2000) has analyzed leaving home amongst young people in Britain and Spain using a theoretical framework that can be translated into a statistical model. This model identifies factors hypothesized to be important in influencing the process of leaving home; these include the respondent's own characteristics and also the economic resources of the parents (measured by parental social class) and the parents' educational level. The results suggest that cultural differences between the two countries remain important and thus challenge claims of convergence in the attitudes and aspirations of young people in Britain and Spain.

# 3.2.3 Policy transfer

Cross-national comparisons of the kind discussed above, which identify crucial factors in explaining differences in a particular outcome, can often be used to make policy recommendations. For example, much analysis of women's employment patterns across Europe identifies affordable childcare as a necessary (but not sufficient) condition for improving women's access to the labour market. Similarly, analysis of welfare-to-work schemes in the USA has been used to promote their implementation in the UK as a way of getting benefit claimants back to work. This research strategy may be seen as analogous to an experimental research design – but with one important difference. An experimental research design holds constant all other causal factors except that under investigation. It is obviously not possible to impose this kind of control in a social situation. Therefore policy interventions which are imported from other countries often have quite different outcomes from those intended, thus highlighting the need for a careful policy analysis that recognizes other relevant factors that may affect the outcome. Such an analysis may then indicate other interventions that would be needed to achieve the desired outcome.

This discussion has referred to cross-national comparisons in general without reference to specific data sources. In the rest of this chapter we focus on census micro-data specifically. We begin by considering the particular strengths offered by census data and then move on to the various steps necessary in conducting a comparative analysis, before finishing the chapter with a worked example from a study of the USA and Britain.

## 3.3 The role of census microdata in cross-national comparisons

### 3.3.1 Comparability of population base and sampling design

There is likely to be considerable cross-national comparability in the population base and sampling design of samples of microdata drawn from censuses of population. In large part this reflects the fact that, as discussed in Chapter 2, there is no need for the complex design used in surveys in order to maximize accuracy whilst reducing the cost of data collection. The large sample size, extensive geographical coverage and comparability in the population base all facilitate comparative research. Whilst there are differences between countries in the sampling design used (see Appendix 2.1), this will be much less than between most social surveys.

Under-enumeration in censuses will always introduce bias. However, available evidence suggests that the level of under-enumeration is very similar in most western countries – about 2–3 percent – and the characteristics of those missed by the census are also similar – the unemployed, young men, those living in the most urban areas and minority ethnic groups (Simpson, 1999). Thus whilst bias from non-response is always a concern, there is no reason to suppose that it is a particular problem for cross-national comparisons using microdata. Moreover, as much cross-national research will use the country as the unit of comparison, bias associated with differential under-enumeration of small areas will be minimized. It is, however, always important to establish whether under-enumeration is likely to have a different impact between countries due to the nature of the analyses being conducted.

### 3.3.2 Comparability of topics and definitions

There are internationally agreed recommendations and requirements for the topics and definitions used in population censuses which help to provide some consistency between different countries. Efforts to achieve international comparability date from 1872 when the International Statistical Institute met in St Petersburg and agreed on a decennial census with enumeration over a 24-hour period and the need for standard classifications of the information collected (Brown, 1978).

In the 1940s the United Nations published a *Population Census Handbook* (UN, 1949), making general recommendations for the 1950 round of censuses (Brown, 1978). In more recent times the Statistical Office of the United Nations makes regular reports on the *Principles and Recommendations for Population and Housing Censuses* (UN, 1998 is the most recent). For Europe, a Working Group of the UN Economic Commission for Europe makes recommendations for the inclusion of topics (e.g. income and ethnic group) and for the definitions of key concepts such as households.

The Statistical Office of the European Community (EUROSTAT) has required standardized outputs from member states since the 1970 round of censuses. Members of the European Union are required to provide information on an agreed set of topics – although in some cases this is not obtained through questions on the census. For example, although age of dwelling is required, the British Census offices are given a derogation or exemption and do not include this question. Some EU member states do not conduct censuses (e.g. the Netherlands, Ch. 1) and therefore obtain all the data required by EUROSTAT from population registers or from surveys.

However, very few countries (if any) completely adhere to UN or EU recommendations. In the 1990 round of censuses, Britain was the only EU member state to ask about ethnic group and no EU members asked about income. Generally, however, there is a wide range of questions which are comparable across most censuses. Because most of the questions are 'factual' rather than attitudinal, comparability is rather more straightforward – nonetheless, as we shall see below, this cannot be taken for granted.

Even when questions are similar, the definitions used may differ. For example, the USA and British censuses use different definitions of household heads. In the USA, this relates to the householder (one of the household members who owns, is buying or rents the accommodation) whilst the British census asks the form-filler to give the head or joint head of household as the first person on the form, and this person is then taken as the head of household (see Section 5.3). Definitions also change over time and therefore countries which are comparable at one time point might not be comparable at another. This is illustrated by changes in the French census with respect to the definition of an adult child, living with parents, not married and with no child of his or her own. Until 1990, the French census classified any such child of 25 years or more as 'un isole'; from 1990 onwards the age limit for an adult child in a family was removed and the French definition became the same as that used in Britain.

There are also differences in how editing and imputation of missing data are carried out – between countries and over time. Hence, while census data may initially appear comparable, it is always important to check carefully for differences in question wording, whether the wording has the same meaning in the countries concerned, a difference in the definitions used and, of course, differences in editing and imputation.

## Conceptual differences between countries

Although most of the questions asked in censuses are very straightforward – e.g. age, sex and marital status – nonetheless there are some where conceptual differences lead to non-comparability. Questions on ethnic or racial origin provide a good example and are discussed in Section 3.5.

Most questions are much less sensitive than ethnicity or race. However, even with an internationally agreed concept such as economic activity, there may be problems of comparability which derive from the question wording or from national differences in definition. For example, we may find differences in reporting of unemployment and economic inactivity associated with welfare regimes. In countries where the criterion of 'looking for work' is strictly enforced in claims for unemployment benefit, this may influence whether respondents identify themselves as unemployed or economically inactive. In Britain, women have a lower unemployment rate than men, whether based on self-definition in the census or the International Labour Organisation (ILO) definition used in the Labour Force Survey, which contrasts with most other European countries. One reason for this may be that the British benefit system makes it very unlikely that partnered women will be able to claim any benefit after the first 6 months (contributory unemployment benefits lasts for 6 months only, after which individuals are assessed on the basis of their family income). At that point, women may opt out of the labour market or take a low-paid, part-time job in order to generate some income, thus reducing recorded and 'real' levels of unemployment. However, in the USA, women also have lower unemployment rates than men, which cannot be readily explained by the welfare regime and indicates the value of widening the comparison in order to better understand the reason for differences.

## 3.4 Steps in conducting cross-national comparisons using census microdata

In the following sections we go through the stages necessary for making comparisons between one or more countries, beginning with decisions over which countries and ending with methods of analysis.

### 3.4.1 Planning comparative research design

The first stage of comparative research design is to decide which countries or regions to analyze. The answer will largely depend on the research question. Theory testing (Section 3.2.1), to establish the generality of predictions or hypotheses about social relationships, requires data from a large number of countries. The countries chosen will relate to the theoretical predictions but will usually be confined to western industrialized countries, unless the relationship is so fundamental that it is predicted to occur in all countries regardless of the level of industrialization.

Where the aim is to refine and develop theory, a smaller number of countries will be chosen and they will often be selected to provide critical cases. A similar research design would be used to establish whether a theory developed in one national context could be generalized to others. For example, one might test whether the relationship between occupational gender segregation and reduced levels of earnings found in Britain (Hakim, 1998) is found elsewhere. An appropriate country for comparison would be Sweden, where earnings are determined within a more formal bargaining context. In this example, we are selecting critical cases to establish the contexts in which the relationship holds.

Model's (1997) analysis of ethnic labour market segregation is an example of theory specification and refinement (Ch. 8). A theory developed with respect to Blacks in the USA is tested on Black Caribbeans in Britain as a way of specifying the process more exactly. This research design depends on convergence of behaviour in two different settings.

When conducting analyses aimed at understanding national differences (Section 3.2.2) – e.g. why women's part-time working is higher in some countries than others – it is most appropriate to select countries where there are clear differences in the outcome of interest and in specific causal factors (e.g. state provision of childcare) but similarities in contextual factors such as political systems or levels of enrolment in higher education. Comparisons between France and Britain (Glover, 1992), France, the UK and the USA (Duane-Richard, 1998), and the USA and

Japan (Houseman and Osawa, 1998) have been used to address this question. It is important to note that census microdata capture only some of the factors that are likely to influence part-time working. Many other influential factors – the extent of labour market regulation, the ease of access to childcare, fiscal policy, and historical and cultural context – cannot be captured by individual-level data. In these examples, countries were chosen because they were known to have different outcomes (different levels of part-time working). Similarly, Heath and Miret Gamundi (1996) chose Spain and Britain for an analysis of young people's living arrangements on the basis of the differences between the two countries. British census data were used with Spanish survey data to highlight the differences – for young people of similar age and educational background – but explanations had to be sought in factors not measured by the census, in this case the different cultural context of the two countries and the differences associated with family life. A further example using census microdata is given by Boyle *et al.* (1999b) who conducted a comparison between the USA and Britain to ask whether family migration is detrimental to the employment status of female partners. An account of the methodological work needed to establish a comparative basis is given in Section 3.6, where they describe the decisions that they needed to make in order to maximize comparability between their data sources.

## 3.4.2 Locating data sources

Data availability is a common limiting factor in designing comparative research, particularly where national statistical offices place restrictions on the use of census microdata outside the country of origin. Access arrangements also change over time as new legislation or confidentiality restrictions either ease or restrict availability. Some NSIs (Canada, Australia) are willing to sell microdata files to both home and foreign purchasers; the USA has long made microdata files available for academic use free of charge through data archives. Appendix 2.1 gave details of the costs and arrangements for overseas access to the 1990 round of censuses for these countries, and the availability of European data was set out in Table 2.1. However, the conditions of access to census data are likely to change between censuses and researchers need to check the current situation with the NSIs concerned. Where data cannot be allowed outside the country of origin, collaboration with a researcher in another country is often a very effective solution and has the benefit of ensuring a source of expertise on the countries concerned.

A number of projects are bringing together samples of microdata, recoded to achieve international comparability, and made available under special access conditions. The Historical Censuses Project at the University of Minnesota is proposing a decade-long initiative to integrate sample census microdata and documentation worldwide for dissemination over the web (www.ipums.umn.edu). The Integrated Public Use Microdata Series International (IPUMSi) is planning to include data for Canada, Mexico, Brazil and Australia, along with that for the USA. This project is designed to provide academic access only with a high degree of confidentiality and usage safeguards built in. Negotiations with the UK are underway and will probably result in the integrated UK files being available to overseas academics once appropriate licences and registration have been completed.

The United Nations Economic Commission for Europe has established a databank of census-based samples as part of a project on population ageing carried out by the Population Activities Unit, based in Geneva. This brings together data files, recoded and standardized to give international comparability, from a growing number of countries in the western world.

## 3.4.3 Deriving comparable datasets

Having obtained the relevant microdata and decided on the hypotheses to be tested and the analytic strategy to be followed, the next stage involves setting up comparable datasets. This can be broken down into a number of steps, as described below.

### Selecting comparable populations

- Establish the coverage and years of the census for the relevant countries. There are often differences in how non-residents are treated, or whether the institutional population is included in a census. For example, until 1991 the Canadian census omitted non-residents, whilst the British census has, since 1961, enumerated everyone present on census night, or usually resident in the household, irrespective of their legal status. (However, the 2001 British Census will be restricted to usual residents.)
- Establish the sampling strategy used for the microdata files and whether this will lead to any problems in comparability.
- Decide on the sample to be included in the analysis – e.g. whether it is to be restricted to usual residents; whether it should include the institutional population. Analyses may also be restricted to adults, to those of working age, or to the elderly. This may appear straightforward, as biological age is equivalent in different countries. However, if age is being used to refer to distinct stages in the life course then the same biological age may not be equivalent in terms of labour market transitions such as leaving school or retirement. This is discussed in more detail below. If the analysis is based on household samples, then it is important that family or household units are comparable. For example, Boyle *et al.* (Section 3.6) discuss the difficulties caused by an inability to identify concealed cohabiting couples in the US PUMS when making comparisons with British microdata.

### Deriving comparable variables

- Ideally we wish to maximize the comparability of the key variables used in the analysis. As suggested above, this may require deriving variables (Ch. 5) in order to achieve functional equivalence. Analyses of economic activity or unemployment will usually be restricted to the population of working age, but this will vary between countries – at the older end of the spectrum in terms of state retirement age and at the younger end in terms of school leaving age. Therefore comparable variables may be derived to represent these transition points. Similarly, analysis of the relationship between family formation and women's labour force participation may distinguish women with pre-school children and school-aged children rather than use the biological age of the child. As the age at which children start school varies considerably between countries, the definition of this variable will also vary.

- Deriving comparable variables for economic activity or unemployment requires considerable thought over how to maximize functional comparability and will invariably be improved by discussions with informed researchers in the countries concerned. Sometimes it will be possible to derive a standard measure, e.g. the ILO unemployment measure. But often census data does not include sufficient detail for this to be derived. In Britain, decisions have to be taken over whether young people on government training schemes should be classified as in employment or unemployed. This is complicated by the fact that the definition of unemployment used by government statistics has changed repeatedly over time (Levitas, 1996).

- There are now a number of international classifications for occupation and education which facilitate international comparisons and these can be added to microdata files if they are not already present. The British SARs contain the International Standard Classification of Occupations (ISCO) and several measures of social standing developed to be internationally comparative, described in Section 2.5.6.

  However, even when these variables are available they need to be used with care. They are usually derived from nationally specific classifications which may use definitions which are not comparable between countries. Thus the definition of a manager is substantially broader in the Britain Standard Occupational Classification than in the classification used by most other European countries, (Dale and Glover, 1990; Birch and Elias, 1997). This leads to non-comparability between countries which cannot always be overcome by recoding to a common international classification.

- It is notoriously hard to achieve comparability of educational qualifications. These may be measured using criteria such as 'age left school' or 'highest qualification obtained', but it is difficult to establish equivalence between countries. For example, many countries have schemes for vocational training and, without detailed knowledge of how these operate, it is difficult to code them to a hierarchical classification. In her comparison of the USA and Hungary, Egerton (1993) distinguished those with higher or tertiary education from those without any tertiary education. The latter group was not disaggregated further – thereby overcoming some of the difficulties confronted by Model (Ch. 8) who set up three levels of post-secondary education to make comparisons between the USA and Britain.

## 3.4.4 Strategies for analysis

### Comparing tables

Much comparative work is based on very simple tabulations. These may be drawn from published reports (e.g. Eurostat), but the ability to use microdata enables the analysis to maximize comparability, in terms of the population base and definitions used, but also by introducing appropriate control variables.

When one is able to produce a set of tables based on variables which have been harmonized for each country, loglinear models may be used to provide a formal way of testing for country differences. Thus one might model the relative chances of reaching a professional or managerial social class position for men and women with different educational qualifications and test the hypothesis that the observed relationship was the same across two different countries. If the hypothesis cannot be rejected then one can assume that the process of using qualifications to obtain occupational advancement is not statistically different in the two countries. Similarly, linear regression models can be used where there is an interval-level response variable, e.g. income; the fit of models with and without interaction terms for country can be used to establish whether there is a difference between the countries in the process of income attainment. These methods are discussed further in Chapter 7.

Aggregate data are also used to make comparisons of occupational segregation, either across country or across time. If these depend on a single summary value for each year or each country, they may produce misleading results because of differences in the size of the marginal values for each year or each country. There has been considerable debate over how best to measure occupational segregation (Hakim, 1992, 1993; Watts, 1993, 1994, 1997; Blackburn *et al.*, 1995; Siltanen *et al.*, 1995) and, indeed, whether a single summary measure is the most appropriate (Hakim, 1998).

However, Siltanen *et al.* (1995) provide a detailed discussion of the methodological strengths and weaknesses of a range of summary measures of segregation. They offer a detailed step-by-step guide to applying a technique known as marginal matching that produces a statistic, Kendall's $tau_b$, which can provide a basis for comparison across countries. The value of this method is that the measure has the following attributes:

- symmetrical with regard to men and women
- constant upper limit indicating total segregation
- constant lower limit indicating no segregation
- size-invariant
- has occupational equivalence
- is sex composition-invariant
- is occupation-invariant.

This therefore provides a basis for comparison of segregation which avoids introducing artefactual effects related to the different size of categories. The same technique can be extended to cross-national comparisons of the extent of occupational segregation for women from different ethnic groups.

## *Multivariate analysis: setting up models for comparative purposes*

One of the key reasons for using microdata is that it allows multivariate analysis (Ch. 7). In this, case decisions must be taken as to whether to combine datasets from each country and to analyze the data jointly or to apply the analytical model to each country independently.

The decision will depend on the research question. If it is assumed that the underlying process is the same for each country, datasets can be combined and the model will highlight the differences in the outcome variable for each country, controlling for other explanatory variables. For example, in a model to predict unemployment in Britain and the USA, the results would indicate whether the chance of unemployment was higher in one country or the other, holding constant all other terms in the model.

This approach will therefore be best suited if the analysis is designed to test a general model in the two countries. The main effects model fitted will be based on the assumption that the effect of other variables, such as sex, is the same in each country. If the theoretical basis suggests that this is not appropriate, then interaction terms can be included in the model to test for inter-country differences.

However, if the analysis is explicitly designed to explore these inter-country differences in more detail, particularly if the theoretical approach is to establish the impact of different social, cultural and political contexts, then it is more appropriate to apply the model to each country in turn. In this case, however, comparisons between models need to be made with care. If using ordinary least-squares regression (Ch. 7), the constant will be different between models and therefore one needs to take this into account when interpreting the effect of coefficients. Similarly, with logistic regression, the reference categories may have different probabilities – it is therefore important to choose reference categories which are as similar as possible with regard to the probability of the outcome variable. (The modelling techniques referred to here are discussed in detail in Ch. 7.)

In earlier chapters we referred to the value of census microdata in the analysis of small population groups, particularly minority ethnic groups, and in Chapter 8 we provide examples of international comparisons of minority ethnic groups. However, this is an area where conceptual differences between countries and between censuses require careful attention. The following section explores this topic in more depth.

## 3.5 The development of questions on race and ethnic identity in US, Canadian and British censuses

Earlier sections referred to the difficulties of making cross-national comparisons in topics where the underlying concept differed between countries. This is illustrated most clearly by questions on race or ethnic identity. In this section we review the development of questions on race or ethnic origin in the USA, Canada and Britain.

The inclusion of questions on ethnic origins of populations has frequently raised problems for both government officials responsible for designing questions and researchers analyzing census returns. All questions included on a census form must be unambiguous in meaning and acceptable to all members of society if the census exercise is to be successfully carried out. Of all of the questions routinely included in censuses, those on ethnic origin frequently pose the biggest problems, particularly when designing a question that can effectively capture the ethnic composition of a population while remaining acceptable to all groups. For this reason only a handful of countries have successfully included questions on ethnic group in census schedules. For example, plans to include an ethnic group question in the 1981 British Census were dropped because of concerns raised by both ethnic minority communities and government that it would increase non-response (Bulmer, 1986). In other countries, such as France, the inclusion of a question on ethnic origin would be unconstitutional (Rachedi, 1994; Coleman and Salt, 1996).

The design of an ethnic origin question must overcome a number of problems and considerations. For instance, ambiguity over definitions of 'ethnic origin' may lead to conflicting responses as to what information the census is trying to collect and the information provided by respondents. Self-identification to an ethnic group is a complex process, reflecting place of birth, ancestry, race, religion, language etc., yet the design of the ethnic question for a census must remain straightforward and concise and cannot necessarily reflect diverse definitions of ethnic origin. Where questions on ethnic group are included in a census, the rationale and structure for these questions will closely reflect country-specific criteria, including the history of the formation of ethnic communities (usually recent immigrants, though not necessarily, as in the case of North American Indians) and the purpose for which the data will be used, such as to identify racial discrimination. Moreover, ethnic origin is not a fixed identity, and an individual's membership of an ethnic group may develop and change over time, reflecting, for example, change of religious affiliation or marriage into another ethnic group. The ability to afford respondents sufficient choice to adequately reflect the ethnic composition of a country must be reconciled with the needs of the census agency to generate concise groupings reflecting the major ethnic groups. The definition of ethnic or racial categories is strongly influenced not only by national contexts but also by the political struggles of the groups themselves. This leads to considerable difficulties in maintaining consistent classifications over time, as ethnic or racial boundaries are reconstructed and preferred terminology changes.

While the inclusion of questions on ethnic origin/cultural identity in censuses might suggest that there is a general public acceptance of what constitutes ethnic origin and harmonization in the way this information is collected, a review of the questions asked between countries and within countries over time illustrates that this is not the case. In general there is little consistency in the questions asked in USA, Canada and Great Britain, and in each country questions on ethnic origin have been subject to major revisions over relatively short time periods. In particular, various questions asked on 'ethnic background' reflect a complex composition of ethnic, racial, religious and linguistic characteristics. In the following sections, we review briefly the development of questions on ethnic origin in each of the three countries and highlight the major differences in the countries concerned.

## United States

Of all three countries, the development of ethnic origin questions in the USA most closely reflects racial, as opposed to 'ethnic', characteristics of the population. The first American census in 1790 classified the population as free or slaves and, if free, white or non-white (Waters, 1990; Lee, 1993). In all subsequent censuses, questions have been included on racial classifications, although the classifications used have changed markedly over time, reflecting both political factors as well as the racial/ethnic mix of the population. One of the first additions to the white/black division was the inclusion of 'mulatto' and 'quadroon' groups, reflecting a preoccupation in distinguishing Black, partly Black and White populations (Lee, 1993). However, by 1900, mulatto and quadroon were dropped and replaced by a Black/White dichotomy. This racial distinction was based on the 'one-drop' principle, whereby any non-White ancestry (primarily Black), no matter how small, legally defines an individual as non-White (Norment, 1995). Throughout the twentieth century, White and Black community groups alike have supported this one-drop policy. Opponents of more recent attempts to recognize mixed race groups state that it will generate an unnecessary plurality of racial groups, which will effectively 'divide and dilute Black political power' (Norment, 1995). However, this principle has become outdated as inter-ethnic and inter-racial marriages in the USA are creating a more diverse racial and ethnic population. There has, therefore, been much discussion in the USA over the recognition of a mixed race category and its inclusion in the 2000 Census (Townsel, 1996; Pearlmann, 1997). In recognition of the difficulty of restricting respondents of mixed racial origin to one racial category, the US Bureau of the Census has decided that, for the first time since 1890, respondents will be able to identify themselves as mixed-race. This will be done by means of multiple responses to a single question and not a 'multiracial' category. This highlights both the politicized and contested nature of the concepts of 'race' and 'ethnicity' and their fluidity over time.

However, the development of a racial classification in the US Census is by no means restricted to the Black/White dichotomy. During the twentieth century, additions to the 'racial' classification have included Chinese, Japanese and Indian (1890), Mexican, Filipino, Hindu and Korean (1930), Hawaiian, part-Hawaiian, Aleut and Eskimo (1960), Vietnamese, Asian Indian, Samoan (1980), leading to the full classification used in 1990 (Lee, 1993). The development of this racial classification over time highlights inconsistencies in the definitions used. For example, while the Black/White division was strictly adhered to, other non-Whites have been granted greater opportunities for self-identification, such the early inclusion of Chinese and Japanese 'racial' groups. These groupings reflect a further ambiguity as they incorporate country of origin and ethnic affiliation as much as they do a racial category. The inclusion of Hindu (a religious denomination) in 1930 and 1940 also highlights the confusion in the underlying definitions used to recognize 'racial' groupings.

In summary, the US racial classification has been primarily concerned with identifying White/Black populations and a small number of additional groups reflecting other visible ethnic communities. The one group not identified, except in 1930, is the Hispanic community. Rather than recognizing the Hispanic community as a 'racial grouping', questions on Hispanic origin have been included as a separate ethnic question, a status not afforded to any other ethnic group. A further extension to questions on ethnic origin was introduced in 1980, with a question on ancestry, thus allowing the majority White population to identify its ethnic origin. Until 1980, the ethnic mix of the White population was only identifiable through questions on country of birth. The question on ancestry was therefore included to identify second and later generation migrants. This differs from previous questions on race in that it explicitly allows respondents to identify two ancestries and hence gives detailed information on the ethnic mix of the population.

## Canada

A question on ethnic origin has been included in Canadian censuses since 1871. Originally the question was aimed at quantifying the British/French mix, although over time questions on other ethnic groups (primarily European) and Native Americans have been included, to give four main ethnic groups (British, French, other European, Aboriginal). Until the 1980s, the design of the ethnic question incorporated two underlying principles. First, that origin should refer to paternal origin rather than self-identification with an ethnic origin, with paternal origin defined as origin of the respondent or respondent's ancestors on first coming to North America. Second, that no allowance was made for mixed origins (hence the choice of 'paternal origin'). As Kralt (1990, p. 27) describes, the impact of the approach was that:

> By forcing all Canadians to report only one ancestry or ethnic origin in the censuses prior to 1981, it was relatively easy for both the researcher and the general public to paint a picture of a Canadian society that consisted of four large, generally mutually exclusive ethnic or cultural groups ... this rather static and simplistic view of Canadian Society was and is simply not valid.

By the 1980s, the introduction of self-completed questionnaires and computer coding facilitated the introduction of an ethnic origin question that would more accurately reflect the ethnic composition of Canada. In the 1986 Census, the two underlying principles had been dropped, and no reference was made to a respondent's origin on first arriving in North America (in recognition that this was offensive to aboriginal populations), although the question still referred to ancestors' origins. Furthermore, respondents were encouraged to tick as many ethnic origins as applied (Kralt, 1990). In 1996, the ethnic origin question was radically altered in that no pre-listed categories were given; instead respondents were provided with four blank spaces in which to write in their origins. This change effectively means that it is not meaningful to compare the ethnic composition recorded in 1996 with previous census years (*The Daily*[1]).

In the 1990s, further developments to census questions on ethnic origin have concentrated on distinguishing between visible and non-visible ethnic groups. 'Black' was included as an ethnic origin in the 1986 Census, along with other groups such as Jewish and Ukrainian, which reflect similar ambiguities in identifying what is meant by ethnic origin discussed in the American case above. In 1996 a new census question was introduced which asked respondents if they belonged to one of the visible minority population groups recognized under the provision of the Employment Equity Act (*The Daily*). These minority populations are: Chinese, South Asian, Black, Arab/West Asian, Filipino, Latin American, Southeast Asian, Japanese and Korean. These two questions (on origin and membership of visible population) incorporate a similar approach to the race and ancestry questions asked in the US census. Finally, an *identity* question was included on aboriginal identity. Again it is difficult to compare the results of this question with those recorded in previous censuses where aboriginal identity was established by the question on ethnic origin.

## *Britain*

A question on ethnic origin was included in the British census for the first time in 1991. As mentioned above, proposals to include a question in the 1981 Census were dropped after concerns were raised that it might adversely affect response. The 1991 British Census Office specifically tried to avoid asking about *racial* group as this was considered politically sensitive and likely to give the appearance that the census distinguished people by race, or skin colour, alone. The question therefore referred to 'ethnic group' although the categories largely reflect racial differences and there are no distinctions within the 'White' population (95 percent of the population)

---

[1] *The Daily* is sent to every MP and Senator informing them of the results of every data release.

which contains considerable ethnic diversity. This lack of differentiation within the White group has been the subject of much criticism (Anthias and Yuval-Davis, 1992; Ballard and Kalra, 1994). Respondents were, however, allowed to give a mixed-ethnic group response by writing in their 'ancestry'. In 1991, about 55 000 respondents described themselves as mixed Black/White. However, a 'mixed' ethnic group raises considerable conceptual difficulties and suggests that respondents were in effect identifying the fact that they were of mixed race. The White Paper announcing plans for the 2001 British Census (HM Government, 1999) suggests a modification of the 1991 question to specifically identify 'mixed' groups, but refers to cultural identity rather than race. It also introduces Irish as an additional sub-category within 'White'. This latter recommendation follows intense lobbying by Irish groups who argued that the Irish were subject to discrimination and disadvantage and should therefore be identified separately.

This brief review of census questions on ethnic origins illustrates how these questions are subject to change over relatively short periods of time. Even where changes may appear relatively subtle, researchers must also be aware that wider social and political change may have an impact on the way in which respondents record their ethnic origin/identity/race etc. Care must therefore be taken when comparing groups over time as well as between countries. The difficulties encountered in making comparisons between ethnic or racial groups in the USA and Britain are discussed more fully by Holdsworth in Chapter 8.

## 3.6 Integrating the SAR and PUMS: a cross-national microdata study of the effect of family migration on women's employment status

*Paul Boyle, Thomas J. Cooke, Keith Halfacree and Darren Smith*

The following section provides a case study of a comparative analysis based on US and British census microdata. The authors provide a detailed account of the issues which they confronted in setting up the research, the decisions that were taken and an indication of some of the results.

There is a growing literature that considers whether family migration has a negative impact on the employment status of women. This is based on the assumption that many families migrate to follow the male 'breadwinner' – with the result that the female partner becomes a 'tied migrant'. This question remains unanswered, partly because studies have provided conflicting evidence. While Boyle and Halfacree (1996) identify significant differences in the employment status of married men and women after family migration, Cooke and Bailey (1996) show that many women's careers may actually benefit from family mobility. Unfortunately, the majority of such studies are geographically specific – Boyle and Halfacree (1996) considered South-East England and Cooke and Bailey (1996) focused on the American mid-West – and it seems unlikely that the results from either are generalizable.

This project links the GB Sample of Anonymised Records (SAR) for 1991 and the US Public Use Microdata Sample (PUMS) for 1990 to investigate this problem for both nations in combination. This is not the first study to link these two datasets (e.g. Model, 1997) but we maintain that it is the first to scrutinize the household and family structures within these samples in such depth.

The two samples vary in a number of ways. Briefly, more questions are asked in the US census, the PUMS is based on a larger sample of census returns than the SAR, and the SAR is more 'refined' than the PUMS, which is relatively 'raw'. For example, the 35 categories of ethnic group coded on the GB census database have been collapsed into 10 categories in the SAR. A surrogate ethnicity variable can only be derived from the PUMS based upon the conflation of race, Hispanic status and ancestry variables, but all three of these variables are provided in the PUMS in full. While the SAR is easier to use, therefore, the PUMS includes more detailed information, allowing experienced users to define their own variables more easily. Our aim, then, was to test whether family migration is detrimental to the employment status of female partners using national-level data.

The first problem was deciding how to extract a matching sample of partners (married or cohabiting) from both datasets and the second problem was generating comparable variables from both sources to investigate the employment status of (non)migrants. Differences between the datasets made this more difficult than might be imagined at first sight and our investigation suggests that, while the benefits of cross-national comparison are undoubted, studies of relatively simple questions, such as the link between family migration and female employment status, require careful and time-consuming data manipulation. Compromises are necessarily involved, and Boyle *et al.* (1999a) provide a more detailed discussion of some of the difficulties encountered.

## 3.6.1 Identifying matched couples

Our analysis required the extraction of married or cohabiting partners aged between 18 and 59, excluding couples where one or both partners were: outside this age group, a student, permanently sick or a member of the armed forces. The individuals in both the PUMS and the SAR household file are grouped into households but we were concerned to identify partners within families, rather than households – couples living within shared rental accommodation might not be related to the household head, for example. While this was accomplished relatively simply in the SAR – families are numbered within households (FAMNUM) and a derived variable (MHUPOS) identifies linked partners (Holdsworth 1995)[2] – this was more difficult to achieve in the PUMS since only the relationship between the household head and the remaining individuals in a household is specified. A household-level variable (RSUBFAM) delimits the number of subfamilies in a household, and individual-level variables (SUBFAM1 and SUBFAM2) define membership within subfamilies. However, each of the subfamilies is defined only for individuals who are related to the

[2] Chapter 5 provides step-by-step guidance on deriving variables using information about household and family members.

household head, through descent, marriage or adoption, or likewise to the household head's spouse. Also, a subfamily can only consist of a married couple (with or without children) or a single parent with children. We cannot identify either a married couple living with an unrelated household head or any cohabiting couple living with any type of household head – either related or unrelated. It is not possible to identify with accuracy the structure of many multi-family households in the PUMS. We therefore focused entirely on 'nuclear family households' (married or cohabiting partners) in our cross-national samples. Analyses of other family types could only be undertaken for GB separately (see Boyle *et al.* 1999b).

## 3.6.2 Identifying matched variables

We also extracted numerous variables for each individual in our sample. Quantitative analyses of tied migration usually define employment status as the dependent variable in a model that includes a range of explanatory variables, and critical among these was the sex of individuals and whether they were long-distance migrants.

### *Employment status*

We required a comparable measure of employment based upon full- and part-time employment in contrast to unemployment or economic inactivity. The primary economic position variable (ECONPRIM) in the SAR could be used for this, as it distinguishes self-defined full-time (more than 30 hours a week) and part-time employment, with reference to the week prior to the census. The hours usually worked in *any* week are also recorded in a separate variable (HOURS) in the SAR. Perhaps surprisingly, the number of part-time workers (defined in ECONPRIM) who usually worked more than 30 hours in each week (defined in HOURS) was considerable, as was the number of full-time workers who recorded themselves as working 30 hours or less in a usual week. This, and the fact that the full-time/part-time distinction is not provided for the self-employed, means that for some of our cross-national analyses we relied on usual hours worked in any week for GB and usual hours worked in 1989 (HOUR89) for the USA. These are not identical, but seem close enough to produce reliable results. Table 3.1 contrasts employment status for GB and the USA and, although part of the difference will be caused by the slightly different definitions of hours worked, we would also expect genuine gender differences in employment status between the two nations. While American women are more likely to be full-time and less likely to be part-time than British women (see also Scott and Duncombe, 1991), the percentage unemployed were nearly identical according to the one week definition.

### *Migrant status*

Another important issue was the definition of migrant status. Because censuses do not include information about the reasons why people move, it is usually assumed that long-distance migrants are more likely to be moving for employment-related reasons, while short-distance movers are more likely to be motivated by residential requirements. Previous quantitative studies of 'tied migration' have simply compared model parameters for male and female long-distance migrants with non-migrants and short-distance migrants. No previous quantitative study has linked partners

**Table 3.1** Defining employment status and migration (column per cent)

| | GB | | USA | |
|---|---|---|---|---|
| | Male (%) | Female (%) | Male (%) | Female (%) |
| **EMPLOYMENT STATUS** | | | | |
| *Hours worked 1 week prior to census* | | | | |
| Full-time | 73.3 | 34.2 | 76.7 | 47.3 |
| Part-time | 1.0 | 30.5 | 3.7 | 16.5 |
| Self-employed | 17.6 | 4.9 | 14.0 | 5.9 |
| Unemployed/economically inactive | 8.1 | 30.4 | 5.5 | 30.3 |
| Total | 100 | 100 | 100 | 100 |
| | | | | |
| *Usual hours worked* | | | | |
| Full-time | 87.0 | 34.6 | 90.6 | 51.2 |
| Part-time | 2.2 | 33.4 | 3.9 | 18.5 |
| Unemployed/economically inactive | 8.1 | 30.4 | 5.5 | 30.3 |
| Not stated | 2.7 | 1.7 | n.a. | n.a. |
| Total | 100 | 100 | 100 | 100 |
| | | | | |
| **MIGRATION** | | | | |
| *Household head – 12 months (GB); 15 months (USA)* | | | | |
| Migrant | 9.9 | 9.9 | 19.7 | 19.7 |
| Non-migrant | 90.1 | 90.1 | 80.3 | 80.3 |
| Total | 100 | 100 | 100 | 100 |
| | | | | |
| *Individual – 12 months (GB): 60 months (USA)* | | | | |
| 50 km cut-off | | | | |
| Long-distance migrant | 1.3 | 1.4 | 13.8 | 14.0 |
| Short-distance migrant | 8.3 | 8.3 | 33.7 | 33.3 |
| Non-migrant | 89.7 | 89.6 | 51.1 | 51.2 |
| Not stated | 0.4 | 0.4 | n.a. | n.a. |
| International migrant | 0.4 | 0.4 | 1.4 | 1.5 |
| Total | 100 | 100 | 100 | 100 |
| | | | | |
| 500 km cut-off | | | | |
| Long-distance migrant | n.a. | n.a. | 6.0 | 6.1 |
| Short-distance migrant | n.a. | n.a. | 41.1 | 41.2 |
| Non-migrant | n.a. | n.a. | 51.1 | 51.2 |
| International migrant | n.a. | n.a. | 1.4 | 1.5 |
| Total | | | 100 | 100 |
| | | | | |
| *Partners – 12 months (GB); 60 months (USA)* | | | | |
| 50 km cut-off | | | | |
| 2 × Long-distance migrants together | 1.1 | 1.1 | 10.1 | 10.1 |
| 2 × Long-distance migrants joining | 0.0 | 0.0 | 1.5 | 1.5 |
| 1 × Long-distance and 1 × short-distance | 0.2 | 0.2 | 3.5 | 3.5 |
| 2 × Short-distance migrants together | 6.5 | 6.5 | 27.6 | 27.6 |
| 2 × Short-distance migrants joining | 0.9 | 0.9 | 2.1 | 2.1 |
| 1 × Long-distance and 1 × non-migrant | 0.2 | 0.2 | 0.9 | 0.9 |
| 1 × Short-distance and 1 × non-migrant | 1.3 | 1.3 | 3.7 | 3.7 |
| 2 × Non-migrants | 88.7 | 88.7 | 48.7 | 48.7 |
| Other[a] | 1.0 | 1.0 | 1.8 | 1.8 |
| Total | 100 | 100 | 100 | 100 |

**Table 3.1** Continued

| | GB | | USA | |
|---|---|---|---|---|
| | Male (%) | Female (%) | Male (%) | Female (%) |
| *Partners – 12 months (GB); 60 months (USA)* | | | | |
| 500 km cut-off | | | | |
| 2 × Long-distance migrants together | n.a. | n.a. | 4.4 | 4.4 |
| 2 × Long-distance migrants joining | n.a. | n.a. | 0.6 | 0.6 |
| 1 × Long-distance and 1 × short-distance | n.a. | n.a. | 1.8 | 1.8 |
| 2 × Short-distance migrants together | n.a. | n.a. | 33.3 | 33.3 |
| 2 × Short-distance migrants joining | n.a. | n.a. | 4.8 | 4.8 |
| 1 × Long-distance and 1 × non-migrant | n.a. | n.a. | 0.3 | 0.3 |
| 1 × Short-distance and 1 × non-migrant | n.a. | n.a. | 4.3 | 4.3 |
| 2 × Non-migrants | n.a. | n.a. | 48.7 | 48.7 |
| Other[a] | n.a. | n.a. | 4.8 | 4.8 |
| Total | 100 | 100 | 100 | 100 |

[a] 'Other' includes other combinations of non-migrants, short-distance migrants, long-distance migrants, international migrants or migrants with 'unstated' origins.
n.a. = not applicable.

who moved *together*, primarily because such an undertaking is not simple. Even so, this is surprising given that partnered migration is the key process under investigation – but see Hayes and Al-Hamad (1999) for similar work in relation to divorce.

In the SAR, individuals whose address was different at the time of enumeration to that 1 year previously are defined as migrants. The Office for National Statistics (ONS) calculated the distance moved by each individual between the origin and destination ward centroids (these are population-weighted points designed to summarize the location of each ward) and these were then recoded into 13 distance bands for output in the SAR databases. In our analyses, migrant partners classified within the same distance band were regarded as having moved together. It was therefore possible to separate out long-distance migrant individuals who had moved a different distance from their partner (Table 3.1). In previous studies, all long-distance migrants will have been grouped together regardless of whether they actually moved with their partner.

The distance moved is not provided in the 5 per cent PUMS, although it can be calculated as we know the place of residence and place of origin for migrants based on 1713 (continental) public use microdata areas (PUMAs). By comparison, in the 1 per cent Household SAR we only have 12 origin and destination regions. Population-weighted centroids were calculated for each PUMA, based on populations and centroids provided for census blocks, which nest neatly into PUMAs. The great circle route distance (this calculation accounts for the curvature of the earth, rather than treating the USA as being on a flat plane and using straight line distances) was then calculated between each pair of PUMAs, providing a more precise distance measure than that available in the SAR. Linked partners who moved between the same pair of PUMAs were assumed to have moved together, but there remains a small chance that they could have moved separately.

Unfortunately, however, migration is defined using a 5-year interval in the US Census which is incompatible with the GB 1-year interval, making integrated migration studies more difficult than has perhaps been suggested in the past (e.g. Fotheringham and Pellegrini, 1996). Work by demographers and others suggests that solving the 1-year/5-year problem is difficult, if not impossible (e.g. Courgeau, 1973a,b; Kitsul and Philipov, 1981; Rogerson, 1990), as the rates of migration based on a 5-year definition will not be five times the rates derived from a 1-year question. Because both migration definitions provide transition, rather than event data (Rees, 1977), they ignore multiple moves made by the same individuals during the 1- and 5-year periods and it is well established that some people ('movers') migrate regularly, while others ('stayers') move infrequently. Many of the frequent movers would be captured in both 1-year and 5-year definitions of migration making the number of 1-year movers greater than one-fifth of the number of 5-year movers.

We therefore adopted two strategies. In one set of analyses, we were forced to rely on the different migration time periods but we were at least able to identify long-distance migrants who moved together in both nations. In the second set of analyses, we took advantage of an alternative PUMS variable (YRMOVED) that identifies how long household heads have remained in the same residence. This variable seems to have been ignored by migration researchers in the USA and was not mentioned by Fotheringham and Pellegrini (1996). This is not a migration question – it is designed to provide information about the length of time household heads have lived in the same accommodation, but it provides an innovative means of producing matched migration data for the SAR and the PUMS.

However, there are problems with the YRMOVED variable. First, no information is provided about the previous residential location, including whether the individual may have moved into the accommodation from outside the USA, or the distance moved for internal migrants; we cannot distinguish short- and long-distance movers. Second, it relates only to the household head. Although the household head may not have moved, other household members may have joined the household during this time and they would be treated as non-migrants. On the other hand, we were forced to assume, incorrectly in some cases, that the whole household migrated if the household head was recorded as having moved. The SAR data was manipulated such that both partners were recorded as migrants if the household head had moved within the previous year, resulting in the incorrect coding of some individuals as migrants. This solution is imperfect, but the SAR data for GB can at least be used to compare the results from these data with the results from analyses that define migration 'correctly'. Third, the question asks whether people were resident at that address during calendar years prior to the census. As the US census was carried out in April 1990, the household head can only be defined as a 15-month migrant, rather than a 12-month migrant as in the SAR.

Finally, we needed to decide on a reasonable cut-off to distinguish long-distance migrants from short-distance migrants and non-migrants (Table 3.1). In Britain, 50 km has been used in the past (e.g. Boyle, 1995) and in our sample 1.3 per cent were defined as long-distance migrants, compared with 8.3 per cent as short-distance migrants and 89.6 percent as non-migrants (less than 1 per cent were accounted for by international migrants and migrants with unstated origins). While 50 km may be regarded as a reasonable cut-off for Britain, it is a relatively short distance in the USA, where cities are more sprawling, the potential distances that people can move are greater and cultural perceptions of distance may be different. We therefore decided to use 50 km as a first cut-off in the USA and this defined 13.9 per cent of the sample as long-distance migrants. We then calculated a second distance cut-off at 500 km, based on a comparison of the migration distance distribution for migrants in both samples (Fig. 3.1), and 6.1 per cent of the sample were defined as long-distance movers, 41.4 per cent as short-distance, and 51.1 per cent as non-migrants according to this definition (1.4 per cent were international migrants). It was also possible to compare the percentage of the samples that were defined as long-distance migrant couples, using both distance cut-offs in the USA. In GB, 1.1 per cent of the sample were migrants who moved 50 km or more together, while in the USA the comparable figures were 10.1 per cent (50 km) and 4.4 per cent (500 km). Note also that for those migration variables that dealt with individuals, rather than partners, women were slightly more likely to have migrated long distances than men in both GB and the USA.

## 3.6.3 Empirical results

The modelling work undertaken with these datasets is described elsewhere (Boyle *et al.*, 2000). Obviously, it is impossible to identify the impacts of family migration on women's employment status without controlling for other confounding effects. However, it is useful to demonstrate how our different definitions of these key variables impact upon our key question – are female partners who migrate long distances more likely to be unemployed or economically inactive than other women? Further, does the definition of migration influence these results – are women who move long distances *together* with their partners more likely to experience negative employment consequences than other (non)migrant women?

Some interesting findings are contained in Table 3.2. First, long-distance male and female migrants are both less likely to be employed than short-distance or non-migrants in GB and the USA, regardless of how migration is defined. Second, women who move long distances are less likely to be employed than men who move long distances, in both GB and the USA. Third, the likelihood of unemployment or economic inactivity for men and women in GB and the USA varies depending on the migration definition used. According to the household head definition, there was a greater difference between men (8.0 per cent) and women (35.1 per cent) in the USA than in GB (11.1 and 33.4 per cent, respectively). However, this reversed using the other migration definitions; the figures based on the 50 km cut-off were 10.5 per cent for men and 41.5 per cent for women in GB, compared with 6.5 and 35.6 per cent in the USA. Careful work is needed to identify whether the negative effects of family migration on women's employment participation are greater in GB or the USA. This is critical before we can begin to explain the 'paradox' (Bruegel, 1999) of migration being both 'positive' and 'negative' for women's careers. Fourth, the differences between men and women are more extreme for those who moved long distances *together* than for those who moved long distances individually, in both GB and the USA (note that the differences between men and women are again larger in GB than in the USA). This suggests that by ignoring linked partners, previous studies have underestimated the negative effects of family migration on female participation in the labour market.

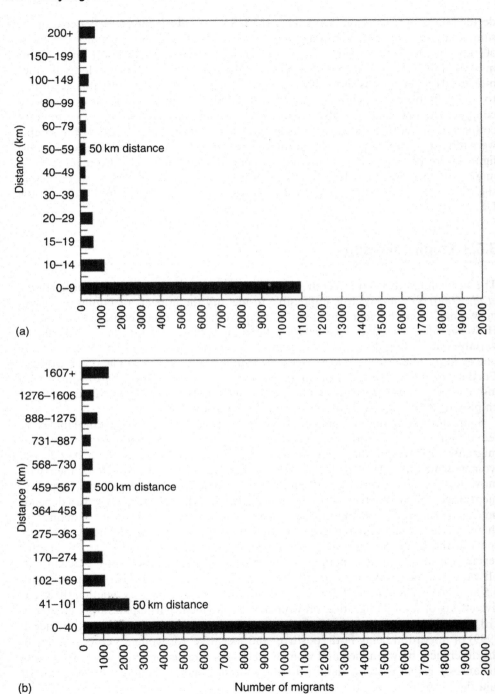

**Fig. 3.1** The distance moved by migrants in (a) GB and (b) the US.

**Table 3.2** Migration by employment status one week prior to census (%)

| | Employment status | | | | | | | |
| --- | --- | --- | --- | --- | --- | --- | --- | --- |
| | GB | | | | USA | | | |
| | Male | | Female | | Male | | Female | |
| | Emp[b] | Not emp | Emp | Not emp | Emp | Not emp | Emp | Not emp |
| **MIGRATION** | | | | | | | | |
| *Household head – 12 months (GB); 15 months (USA)* | | | | | | | | |
| Migrant | 88.9 | 11.1 | 66.6 | 33.4 | 92.0 | 8.0 | 64.9 | 35.1 |
| Non-migrant | 92.2 | 7.8 | 69.9 | 30.1 | 95.1 | 4.9 | 70.8 | 29.2 |
| *Individual – 12 months (GB); 60 months (USA)* | | | | | | | | |
| (50 km cut-off) | | | | | | | | |
| Long-distance migrant | 89.5 | 10.5 | 58.5 | 41.5 | 93.5 | 6.5 | 64.4 | 35.6 |
| Short-distance migrant/non-migrant[a] | 92.0 | 8.0 | 69.9 | 30.1 | 94.7 | 5.3 | 70.9 | 29.1 |
| International migrant | 83.1 | 16.9 | 50.7 | 49.3 | 90.1 | 9.9 | 50.1 | 49.9 |
| (500 km cut-off) | | | | | | | | |
| Long-distance migrant | n.a. | n.a. | n.a. | n.a. | 93.2 | 6.8 | 62.8 | 37.2 |
| Short-distance migrant/non-migrant[a] | n.a. | n.a. | n.a. | n.a. | 94.6 | 5.4 | 70.4 | 29.6 |
| International migrant | n.a. | n.a. | n.a. | n.a. | 90.1 | 9.9 | 50.1 | 49.9 |
| *Partners – 12 months (GB); 60 months (USA)* | | | | | | | | |
| (50 km cut-off) | | | | | | | | |
| 2 × Long-distance migrants moving together | 91.3 | 8.7 | 52.5 | 47.5 | 93.7 | 6.3 | 61.6 | 38.4 |
| Others | 92.0 | 8.0 | 69.9 | 30.1 | 94.6 | 5.4 | 70.9 | 29.1 |
| International migrants with 'any partner' | 84.3 | 15.7 | 42.5 | 57.5 | 91.0 | 9.0 | 54.4 | 45.6 |
| (500 km cut-off) | | | | | | | | |
| 2 × Long-distance migrants moving together | n.a. | n.a. | n.a. | n.a. | 93.8 | 6.2 | 60.4 | 39.6 |
| Others | n.a. | n.a. | n.a. | n.a. | 94.6 | 5.4 | 70.4 | 29.6 |
| International migrants with 'any partner' | n.a. | n.a. | n.a. | n.a. | 91.0 | 9.0 | 54.4 | 45.6 |

[a] Includes migrants with 'unstated' origins in GB.
[b] Emp = employed (full- or part-time); Not emp = Not employed (unemployed or economically inactive).

## 3.6.4 Discussion

Identifying linked partners and matching definitions of employment status and migration for the SAR and PUMS are not simple tasks. Matching samples requires careful consideration of definitional issues and these results show that the decisions taken may have a considerable influence on the conclusions drawn. The PUMS provide far more detail about individuals, partly because the USA census includes more questions, but also because confidentiality constraints imposed upon the data are less strict. On the other hand, careful preliminary work undertaken by the ONS, and furthered by the Census Microdata Unit, has resulted in a more accessible dataset, enabling, and encouraging, non-specialists to make use of this unparalleled microdata resource. In addition, there are many differences between the SAR and PUMS that we have ignored here. For example, the PUMS provides the respondent's place of birth (based on current region, state or county), allowing the analysis of return and onward migration (Long, 1988) but this information is not provided in the GB census. We also spent some time deciding how to distinguish comparable ethnic minority groups for two very differently constituted nations – it is impossible, for example, to identify Hispanics in the GB census, but they are a considerable minority in the USA. Even so, despite all of these problems, we are confident that our samples are matched closely enough to allow us to provide the first reliably integrated cross-national analysis of 'tied migration'.

## Acknowledgements

This research was funded by the Economic and Social Research Council (grant no. R000237318) and the National Science Foundation (grant no. 9729587). The British census data are Crown copyright and were bought by ESRC and JISC for use in the academic community. The UK Data Archive and Manchester Computing Centre made the British data available to us.

# 4

# Methodological issues in using census microdata

## 4.1 Introduction

This chapter is concerned with basic methodological issues which need to be addressed before embarking upon the analysis of census microdata. We focus particularly on issues that are distinctive to census microdata as opposed to other census outputs or social survey data. The chapter draws examples primarily from the British SARs, although most are applicable to census microdata from other countries. The most important features of census microdata affecting analysis relate to the appropriate selection of the population base, the accuracy of coverage of the census itself, and the sampling design and consequent design effects in microdata files. The chapter goes through each of these topics in turn.

## 4.2 Choice of population bases and units of analysis

In tabulated census outputs, such as the small area statistics (SAS) produced from the UK 1991 Census, or the community profiles from the Australian census, the population base and the units of analysis are predetermined at the point at which the table is specified. For example, a table may include counts of all residents in households, all residents over 16 in households, or all households with residents. With census microdata, however, the user has the flexibility to decide what is the most appropriate population base to be used in an analysis. We begin by looking at the basic parameters that determine what choice of population base is available.

### 4.2.1 The census population base

When a census is taken, decisions are made about who is counted and where they are counted. Because a census of population aims to cover the entire population, it includes people usually omitted from most sample surveys. In particular, it includes people living in institutions, e.g. old people's homes, prisons and military bases. UK censuses have always aimed to make a complete count of everyone in the country on census night – irrespective of whether or not they were normally resident outside the UK. However, the enumerated population varies between countries – in 1991 the Canadian Census included non-permanent residents (e.g. students) for the first time but continued to omit overseas visitors normally resident elsewhere. A first step, therefore, is to clarify *who* is enumerated in a census.

The second decision relates to *where* people are counted. There are two basic methods of enumeration. One is to ask each household or institution to enter on the census form everyone present on census night irrespective of where they usually live. The second is to enter on the form everyone usually resident, irrespective of whether or not they are present at that address on census night (Redfern, 1981). In early censuses, Britain recorded the population present on census night and, from 1931, also asked for the person's usual address. Since 1961, households have additionally been asked to enter anyone usually resident in the household but away on census night. From this information, the 1991 British Census was able to provide counts for both the methods outlined above:

- the population present on census night;
- the usually resident population, distinguishing whether each person was present or absent on census night.

The 2001 UK Census will, however, collect statistics only for usual residents – defined as 'any person who usually lives at the address; or who has no other usual address'. Students and schoolchildren will be enumerated as resident at their term-time address with only basic demographic information collected from their vacation address (if different), whereas in 1991 their parental home was taken to be their usual address.

## 4.2.2 The population base in samples of microdata

The population base included in microdata samples may differ from the census population base – that is, at the point of drawing the microdata sample, decisions may have been taken to omit certain groups. In the 1990 US PUMS the sampling universe was defined as all occupied housing units including all occupants, vacant housing units and group quarters (GQ) persons in the census sample. However, the 3 per cent 'elderly' file is restricted to households with at least one person aged 60 or more (or GQ persons aged 60 or more). In both the Canadian and UK microdata samples for 1991, the Household files are restricted to private households and omit the 'institutional' population. By contrast, the 1991 Australian Household Sample File (HSF) contains a 1 per cent sample of households in occupied private dwellings and all persons counted within each selected dwelling, and a 1 per cent sample of persons in non-private dwellings. More details are given in Appendix 2.1.

The analyst therefore needs to establish what population is included in the microdata file before proceeding to select the population base for the proposed analysis.

## 4.2.3 Population bases for analysis

One of the strengths of microdata is the ability to select the population to be analyzed. In the following sections we discuss some of the choices to be made.

## Institutional population

Often one will want to omit the institutional population from an analysis. However, it is important to recognize that, although the institutional population is relatively small – under 3 per cent in Britain – it is distinctive on a number of dimensions and therefore its inclusion or exclusion can affect the results of analyses as well as estimates of numbers. Living in an old people's home or hospital becomes increasingly likely with advancing age. In Britain in 1991, over 16 per cent of men and 29 per cent of women aged 85 and over were enumerated in an institution. As may be expected, they differ in a number of important ways from the elderly who live in their own homes – most notably in their health status, but also in their marital status. Grundy *et al.*, in Chapter 8, demonstrate that the apparent difference in the health status between elderly married people and those living alone is largely explained by the exclusion of those in institutions.

## Visitors and usual residents

In the UK, the 1991 SARs included those *usually resident* in a household or institution as well as those *present* on census night – who may not necessarily be usually resident. The term 'visitor' is applied to those present on census night but not usually resident. Again, the characteristics of the 'usually resident' and 'visitors' are very different and need to be understood before an informed decision can be made. The base most widely used by UK researchers is the 'usually resident' population living in households. This gives greatest comparability with most household surveys and with most tabular output from the census. Using the 1991 SARs this was achieved by excluding visitors from the analysis. This has the additional benefit that it avoids double counting, as visitors may appear elsewhere as usual residents. For the 2001 Census, details on visitors will not be collected and therefore analysis will be simplified. Another population aspect that may cause problems is the status of students, particularly those in higher education. In many countries it is usual to study at a university or college that is in a different geographical location from the 'home' residence of a parent. The British Census in 1991 requested full-time students to record themselves as 'visitors' at their term-time address and asked their parents to record them as 'usual residents' in the parental home. Analysis of the SARs suggests that not all respondents obeyed this instruction (Heath, 1994). For the 2001 Census, students and schoolchildren will be recorded as usually resident at their term-time address where this is different from their vacation address (Section 4.2.1) and therefore comparisons between 1991 and 2001 will need to take this into consideration. By contrast, the 1991 Canadian Census recorded students once only, at their place of permanent residence. The same practice was followed in 1996 and is planned for 2001.

The key point to make is the need to be aware of the consequences of using any particular population base. Whether or not this will have an impact on the results of the analysis will depend upon the topic of the study. Thus in a study of the living arrangements of young people, it would be crucially important to establish how and where students were being counted. The extent to which young people appear to have left home may vary significantly by how and where they are recorded, and this has important implications for cross-national comparisons – see Chapter 3.

## Population base for particular variables

Not all questions in a census are asked of the entire population. Many questions are asked only of adults (usually defined as aged 16 or over) and others, e.g. occupation, are likely to be asked only of those who have a current job or have held a job at some time previously. The 2001 British Census plans to ask occupational information for current or last job of everyone aged under 75. However, in 1991 this was restricted to those who had held a job in the previous 10 years; and in 1981 this was asked only of the economically active and the retired population. Recent Canadian censuses ask occupational information only of those having held a job in the preceding year, while the Australian census obtains this information only for those who had a job in the week before the census was taken.

Therefore, before using occupational information it is necessary to define the population that is relevant. This is particularly important when making comparisons between years or between countries. Analyses which focus on particular employment status groups – e.g. part-time working or self-employment – may need to restrict analysis to all those currently working in order to achieve comparability. Where there are differences between countries in the population for whom occupational information is available, there will also need to be restrictions to achieve comparability. However, decisions over the appropriate population base can only be made by thinking through the proposed analysis very carefully.

For example, if calculating unemployment rates, it would be necessary to express the number unemployed as a percentage of the population at risk of unemployment. This is usually taken to be all those of 'working age' and is typically restricted to those who are economically active. However, decisions over the definition of 'working age' may not be clear-cut and using age 16–65 will produce a rather different rate of unemployment than using age 18–59.

Alternatively, in an analysis of the relationship between social class (based on occupation) and limiting long-standing illness, the analyst will almost certainly want to extend the population base beyond those currently in employment. The 1991 UK SARs contain occupation (and thus social class) for all those who had held a job in the last 10 years. At least for men one might expect this to provide social class for everyone of working age and within 10 years of retirement. However, there is a substantial proportion of men who do not report an occupation; this proportion increases with age and is strongly biased towards those who report a long-term limiting illness. Exploratory work on the SARs readily identifies this bias. Thus, 12 per cent of men aged 16–74 report a long-term limiting illness, but this rises to 33 per cent amongst men aged 16–74 who do not report a social class.

## Level of analysis

It is equally crucial to select the correct level of analysis. In general, variables measuring individual characteristics are most appropriately analyzed at the individual level (e.g. the percentage of *persons* who are Black Caribbean). Similarly, variables measured at the household level (e.g. housing tenure) are best analyzed at the household level; thus one might analyze tenure in relation to household composition.

The choice of unit of analysis can have an important impact on results. For example, if the average number of persons per household varies according to housing tenure, the percentage of residents who are owner-occupiers will be different from the

**Table 4.1** Housing density expressed as persons per room (ppr) in the GB SARs for 1991

| | Individual file | | Household file | |
|---|---|---|---|---|
| | Households (%) | Individuals (%) | Households (%) | Individuals (%) |
| Up to 0.5 ppr | 63.8 | 46.0 | 63.9 | 46.0 |
| >0.5–0.75 ppr | 20.2 | 27.5 | 20.2 | 27.5 |
| >0.75–1.00 ppr | 13.8 | 21.8 | 13.8 | 21.7 |
| >1.00 ppr | 2.2 | 4.8 | 2.2 | 4.7 |
| | 429 173 | 1 063 646 | 214 369 | 531 170 |

percentage of households with that tenure. Table 4.1 shows the way in which the extent of overcrowding appears to vary depending on whether it is based on individuals or households – for the obvious reason that there are more individuals in overcrowded households. Thus, whilst nearly 5 per cent of individuals are living at a density usually considered overcrowded (more than one person per room), this is the case for only just over 2 per cent of households. At the other end of the distribution, nearly two-thirds of households have densities of only 0.5 persons per room or less by comparison with slightly less than half of all individuals.

### Selecting households
Where the microdata file consists of a sample of individuals, a representative sample of households can be obtained by selecting from the sample all household reference people (heads of household).

Where a sample of households has been drawn, analysis at the household level can be conducted either by using the household record (if a hierarchical database is being used – see Ch. 5) or by selecting only 'heads of household' if the file has been flattened to the individual level. Chapter 5 deals in more detail with the structure of hierarchical data files and the different units of analysis that can be used.

## 4.3 Taking account of error in census microdata

As with any sample survey data, there are two potential sources of error in census microdata. First, there is non-sampling error. This arises from error or inconsistency in the responses recorded on the census schedule, in the coding and data entry operation, and from undercoverage by the census. Second, because census microdata is a sample drawn from the whole population, there are sampling errors. In the following section we discuss ways of dealing with both forms of error (see also Section 1.6).

### 4.3.1 Non-sampling error

#### Imputation of inconsistent and missing responses
Despite the best efforts at checking, not all census schedules contain complete information and, for some individuals, the information recorded is inconsistent (e.g. a person aged 3 and in employment). This is dealt with by an editing and imputation procedure. Precise methods used vary; in Britain in 1991 a 'hot-deck' method was used to impute

missing items of information or to replace 'out-of-range' replies. Hot-deck methods insert a value for the missing category which is taken from another record with the same values on related variables and which is geographically proximate (Mills and Teague, 1991; Dale and Marsh, 1993). The 1990 US PUMS use 'allocation' to replace missing or inconsistent responses with data obtained from other respondents who were comparable on related characteristics. The level of item imputation can give an indication of the overall quality of the data – in Britain in 1991 it was about 2 per cent.

Generally, the process of imputation will improve the quality of data, although sometimes it may disallow 'real' responses, e.g. a married 14-year-old or same-sex couples. However, in certain analyses it may be preferable to exclude imputed cases if the researcher has good reason to believe that their inclusion may distort the analysis. In many microdata files, cases that are imputed for particular variables are marked with a flag variable and can therefore be identified. For example, in the US 1990 Census, missing income data were imputed by age group, although all people aged over 65 were grouped in a single category for this purposes. Hence an analysis of income of individuals aged over 80, for example, will include imputed values based on data from individuals aged over 65 (US Bureau of the Census, 1990). It might be expected that the income of the younger age group (aged 65–79) will be greater than that of the very elderly. In this case, inclusion of imputed cases may inflate values for the latter group. It is often useful to perform a sensitivity analysis, in which imputed cases are first included and then the analysis is repeated excluding them, in order to assess the difference in the results.

In the Australian 1991 Census, there is evidence that the imputation process underestimates ages across all broad age groups, with many 15–19-year-olds, in particular, being given ages less than 15 years. Imputed marital status appears able to correctly classify married people in most cases (over 90 per cent) and a majority of never married people (almost 60 per cent), but incorrectly allocates 'never married' or 'married' status to most non-respondents in the other categories of separated, divorced and widowed (ABS, 1993).

## *Other measurement errors*

In addition to this, measurement errors may be caused by inaccuracies in information supplied by respondents, but which are not picked up by edit checks, and by coding errors introduced during processing. These types of error vary from country to country and from year to year, but can be gauged using post-census validation methods. In Britain, a post-enumeration survey has been conducted after each census since 1971 in order to provide a check on both the coverage and the quality of the census (Britton and Birch, 1985; Wiggins, 1993; Heady et al., 1994). The quality of the census responses were checked by drawing a sample of households which had returned census schedules and by asking the same questions as those in the census but in more detail. The consistency of the response between the census schedule and the interview was then used to give a measure of the accuracy of response to the census. The quality of responses to the questions asked in the 2001 Census in Britain have been established *before* the census as part of the 1999 Census Rehearsal.

In the context of microdata, it is important to focus on gross error rates rather than net error rates. The latter are always smaller than gross rates because of the

**Table 4.2** Gross misclassification rates for 1981 and 1991, Great Britain

| Census variable | Gross misclassification rate[a] | |
|---|---|---|
| | 1981 (%) | 1991 (%) |
| *Household variables* | | |
| Number of rooms | 28.6 | 28.3 |
| Central heating | – | 5.9 |
| Number of cars | 3.5 | 5.4 |
| Tenure | 3.2 | 4.5 |
| Accommodation type | 2.1 | 3.3 |
| Bath or shower | 0.9 | 0.8 |
| Inside WC | 1.3 | 0.7 |
| *Individual variables* | | |
| Economic position | 7.8 | 10.9 |
| Daily journey to work | 8.6 | 8.1 |
| Hours of work | – | 8.0 |
| Higher qualifications | 2.3 | 2.6 |
| Usual address 1 year ago | 1.5 | 1.8 |
| Marital status | 1.3 | 1.6 |
| Relationship to household head | 1.3 | 1.4 |
| Country of birth | – | 1.2 |
| Age band | – | 1.1 |
| Sex | – | 0.6 |

[*Source*: ONS Framework Document, 1996, Table 2.1]
[a] 1981 figures are for England and Wales; 1991 for Great Britain.

'cancelling-out' effect of errors and are often used in reporting aggregate results where the concern is with marginal distributions. The results for the 1991 British Census were discussed in Section 1.6, and Table 4.2 provides details of gross error rates. For the majority of variables, the error rate is less than 2 per cent. The pattern and extent of the inaccuracies described have remained fairly consistent since 1971 and are likely to be similar in the 2001 Census.

In Australia, only five questions were checked in the 1991 post-enumeration survey: sex, age, marital status, birthplace and Aboriginal/Torres Strait Islander origin (ABS, 1993). Age comparisons showed a large number of small errors, with 8 per cent disagreeing by 1 year and 2 per cent by more than 1 year. By 5-year age group, the highest level of disagreement of 3–4 per cent was found in the 40–74 year age groups. About 17 per cent of those reporting being 'separated' or 'divorced' in the census gave a different response in the post-enumeration survey, while country of birth showed relatively few discrepancies except for Yugoslavia and 'Other' categories. For 'Aboriginal' and 'Torres Strait Islander' responses in the census, the agreement was much lower: 15 and 50 per cent, respectively, gave a different response, the majority saying they were non-indigenous in the PES.

These kinds of measurement errors occur in all censuses and there is little that the analyst can do except exercise caution over interpreting results where error levels may be high. One obvious example of this is the use of number of rooms to calculate an indicator of overcrowding. It would be a useful exercise to recalculate this indicator several times, making different adjustments for the under- or overcounting of rooms and to observe what effect this had on the measure of overcrowding.

## Under-enumeration

No census manages to obtain a complete enumeration. Estimates of under-enumeration vary between countries and over time, with levels of about 2–3 per cent in Britain, North America and Australia in the 1990 round of censuses. There are a variety of different methods used for estimating the extent of under-enumeration and these are constantly under review. Both estimates and methods used are discussed in detail in Section 1.6.

If those who did not complete and return census forms were no different from the population as a whole (if they were missing at random), then there would be little problem. However, as Chapter 1 showed, there is substantial evidence that non-respondents differ from respondents in a number of important respects, both in geographical location (the problem tends to be greater in cities) and social and demographic characteristics (young men and ethnic minorities are more likely to be missed).

## Correcting for under-enumeration

The selective nature of the undercount, and the fact that a key role of the census is to provide accurate population counts mean that considerable efforts are taken, not only to identify the problem, but to correct for it. However, correction for under-enumeration may have far-reaching consequences that impact differently on different interests.

Different strategies for correction are used in different countries and these also change over time. In Britain, imputation of missing households was used for the first time with the 1991 Census. Where households were identified by enumerators but no-one in the household returned a schedule, basic information for the household and the individuals within the household was imputed. This generated records for about 1.6 per cent of the total population, termed 'wholly absent households'. These imputed households are not included in the 1991 microdata samples.

The UK Census Offices plan to conduct a One Number Census for the 2001 Census (Brown et al., 1999; ONC website). Adjustments for under-enumeration will be built into the processing of census data. This means that all census outputs will be consistent in terms of their base numbers. It also means that, unlike 1991, microdata files will contain imputed records for individuals and households. It is planned that these records will be flagged so that analysts will be able to omit them if they so wish. However, for those using the hierarchical household file from the UK SARs, imputation of individuals deemed missing from responding households will have implications for the relationship between household members.

## Using weighting factors to adjust microdata files

Where adjustment figures have been calculated (and have not resulted in the addition of imputed individuals or households to the microdata files), they can be used to weight microdata in order to compensate for under-enumeration. Weighting factors for microdata files are available for the USA and Canada, and have also been calculated for the 1991 Individual SAR for Britain. They are not available for Australia, but could readily be derived from the figures of the estimated resident population for states, territories and Australia by age and sex published after each census. The following discussion is based on the use of weighting factors in the 1991 SARs. The general principles, however, apply to other microdata files.

The weighting factors derived for the 2 per cent Individual SAR allow the user to adjust the SAR population to the mid-1991 population estimate for each geographical area (a large local authority or several adjacent smaller areas) by single year of age and sex. This population estimate is the most accurate available and includes those imputed or missed altogether by the census. Weighting factors represent the ratio of usual residents in each SAR area, by age and sex categories, to the corresponding number of persons in the Registrar General's mid-1991 estimate for that area (OPCS, 1993a). By applying population weights, census microdata provide a better representation of the 'true' population. The weights can also be used to gross the sample estimate to a population estimate by multiplying by the sampling proportion (50 in the GB Individual SAR). When applying population weights, it is important to use the same population as that for which weights were estimated. For the SARs this means excluding 'visitors', as the Registrar General's estimates were for usual residents. (Although full-time students were counted at their term-time address in the mid-year population estimates, these were redistributed to the areas of 'usual residence' to recreate the census population base of usual residents.)

Table 4.3 shows the effect of applying population weights to the age structure of Greater London. The extent to which the weights increase the population is shown by the ratio of the weighted to the unweighted population. In all age categories the ratio is greater than 1, and this is especially marked in the 20–29 age group. Re-weighting therefore leads to a relative increase in the proportion of men aged 20–29, because of the large undercount within that group. Although most population weights in the SARs are very close to unity, for some subgroups they were based on very small estimated populations and very small cell counts. This means that, for a few cells, weights have quite large values. It is therefore necessary to be cautious when using weights with very small subgroups, because a large weight may exaggerate an unusual case. For example, in the analysis of occupations of women from ethnic minorities, a sampled woman with a large weighting factor, who had an unusual occupation, would strongly bias the estimate for women from that ethnic group as a whole.

It is also important to note that weighting only corrects for a subset of characteristics; in this example, age, sex and geographical area. There is no additional correction for ethnic group. However, because Black Caribbeans, for example, are slightly over-represented in groups for which there is an undercount (men aged

**Table 4.3** Male age structure – London boroughs (residents)

| Age | Ratio[a] | Unweighted percentage | Weighted percentage |
| --- | --- | --- | --- |
| 0–9 | 1.07 | 14.0 | 13.7 |
| 10–19 | 1.07 | 11.9 | 11.7 |
| 20–29 | 1.21 | 18.6 | 20.6 |
| 30–39 | 1.11 | 15.7 | 16.0 |
| 40–49 | 1.07 | 12.5 | 12.3 |
| 50–59 | 1.04 | 10.3 | 9.8 |
| 60–69 | 1.02 | 9.0 | 8.4 |
| 70–79 | 1.04 | 5.6 | 5.3 |
| Over 80 | 1.05 | 2.2 | 2.1 |

[*Source*: 2 per cent Individual SAR]
[a] Weighted $n$ : unweighted $n$ (average weight).

**Table 4.4** Ethnic composition (men) – London boroughs (residents)

| Ethnic group | Unweighted | Weighted |
|---|---|---|
| White | 79.7 | 79.5 |
| Black Caribbean | 4.0 | 4.1 |
| Black African | 2.3 | 2.4 |
| Black Other | 1.1 | 1.1 |
| Indian | 5.6 | 5.5 |
| Pakistani | 1.2 | 1.2 |
| Bangladeshi | 1.4 | 1.4 |
| Chinese | 0.9 | 0.9 |
| Other – Asian | 1.8 | 1.8 |
| Other – Other | 1.9 | 2.0 |

[*Source*: 2 per cent Individual SAR]

20–29 and 30–40 in urban areas), applying weighting factors boosts the percentage of Black Caribbeans (Table 4.4). This does not affect most ethnic groups as they are not under- or over-represented in those age groups. Unless such clustering exists, weighting cannot improve estimates of any variables other than those on which the weights were based.

# 4.4 Sampling procedures and sampling errors

## Populations and samples

Sometimes in scientific research it is possible to do research on all the individuals or cases one is interested in. For example, the census aims to cover all people resident in the country on a particular night. However, most research is done on a subset of the population. When information is only collected on some cases from the population, this is known as a *sample*. The aim of sampling is to ensure that the cases represent the population from which they are drawn. Random sampling (see below) is an important method for achieving this aim. The term population does not necessarily refer to *all* possible units of analysis, e.g. all residents in a country. Instead the population refers to all the units of analysis about which we wish to generalize. This could be, for example, all persons employed in a particular industry in a particular region.

## Sample design

Perhaps the most important difference between census microdata and other census products is that microdata are invariably based on a random sample of the population enumerated by the census. By taking a sample, a degree of uncertainty or sampling error is introduced. The following sections discuss the way in which sampling error affects the accuracy of estimates and how the size of this can be calculated, the effect of the sampling strategy used and how to use this information in the interpretation of analysis.

Appendix 2.1 provided some background information on the sampling design used in a number of countries. Box 4.1 gives further details of sampling for the 1991 SARs for Britain.

## Box 4.1 Sampling strategy for GB SARs

In 1991 the GB Census was coded in two stages. First, easy-to-code variables such as age, sex and marital status were coded for all forms. Second, a geographically stratified random sample of 10 per cent of forms was selected for further coding of more complex questions such as occupation (Dale, 1993). This 10 per cent sample was then used as the basis for the SARs. The 10 per cent sample excluded imputed 'wholly absent households' because the imputed cases did not contain information on variables such as occupation and educational qualifications. Therefore the imputed cases were also excluded from the SARs.

Households in the 10 per cent sample were ordered geographically and grouped into batches of 10. One household was selected at random from each batch to form the 1 per cent Household SAR. The remaining households were stratified into groups of nine, and two individuals were selected at random from each group to form the 2 per cent Individual SAR, ensuring there was no overlap between the Household File and the Individual File (Campbell *et al.*, 1996). It was possible, however, to sample more than one individual from any one household, which resulted in a degree of clustering of individuals within households. Finally, individuals in communal establishments were stratified into groups of five, and one individual was selected at random from each group.

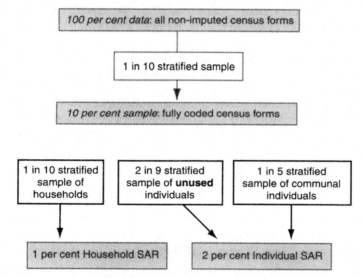

Source: Campbell *et al.*, CCSR Occasional Paper No. 6, 1996. Reproduced with permission from The Cathie Marsh Centre for Census and Research.

## Simple random sampling

Statistical analysis of survey data is normally based on an assumption of simple random sampling whereby units (households or individuals in this discussion) are selected at random from the entire population, each having equal probability of selection, and

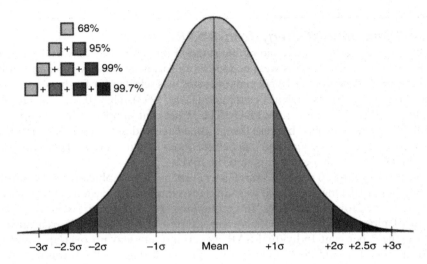

**Fig. 4.1** A normal distribution.

with the selection of one unit not influencing the selection of others. Providing the rules of simple random sampling are followed, then the relationship between the characteristics of the population and the characteristics of the sample can be predicted by probability theory and the properties of the normal or Gaussian distribution.

The *distribution* of a variable is the pattern of values that it takes. The normal distribution is characterized by a bell-shaped curve (Fig. 4.1). If a variable has a normal distribution, it is possible to make inferences about the proportion of cases lying within a particular range of values, based on the area under the normal curve. The properties of the normal distribution allow the area under the curve to be divided proportionately and specified using a measure of dispersion, the standard deviation. Most importantly, 95 per cent of cases fall within approximately two standard deviations of the mean, where the standard deviation ($\sigma$), a measure of variability or spread, is equal to

$$\sigma = \sqrt{\frac{\sum(X - \bar{X})^2}{N - 1}}$$

Note that the use of Greek notation is conventional practice when representing a population parameter. The Roman alphabet is used in equations representing sample statistics (see Box 4.2).

Statistical inference based on assumptions about the distribution of data is known, more generally, as parametric testing. In many instances, the assumptions of parametric tests are not valid, and non-parametric ('distribution free') tests may be more appropriate (see Ch. 6). However, if we are interested in the accuracy of summary statistics (such as the mean) then the assumption of normality is not always necessary.

### Standard errors and confidence intervals

If we take an infinite number of samples for a particular variable, then the mean of a variable such as 'age' will vary from sample to sample and have a distribution known

## Box 4.2 Population parameters and sample estimates

Sample estimates should not be confused with population parameters. A population parameter is the true value of a statistic in the population as a whole and as such is not subject to sampling error but is often unknown. An exception to this would be if you were using 100 per cent census data. In order to distinguish between a population parameter and a sample estimate, a different form of notation is used:

- Greek letters are used to represent population parameters;
- Sample statistics (estimates) are usually represented by the Roman alphabet.

A *sample value* of a variable is given by the letter $x$

The *sample size* is given by $n$

The *sample mean* is given by

$$\bar{x} = \frac{\sum x}{n}$$

where $\sum$ means 'the sum of' and the value of variable $x$, for each case in the sample, is summed and divided by the total number of cases in the sample.

The *sample variance* is given by $s^2$:

$$s^2 = \frac{\sum(x - \bar{x})^2}{n - 1}$$

To calculate the variance

- Subtract the mean from each observation.
- Square each of the differences which result from this subtraction.
- Obtain the sum of squares by adding together all the squared differences.
- Count the number of cases to obtain $n$ and subtract 1 to allow for the fact that a sample is being used.
- Divide the sum of squares by $n - 1$ to give the variance.

The *standard deviation* of the sample, $s$, represents the square root of the variance:

$$s = \sqrt{\frac{\sum(x - \bar{x})^2}{n - 1}}$$

as the sampling distribution of the mean. The central limit theorem tells us that the sampling distribution of the mean taken from a large sample (more than 30) tends to have a normal distribution, regardless of the distribution of the population from which the samples are drawn (Pagano, 1986). Using the properties of the normal distribution, we can calculate that there is a 95 per cent probability that the true population mean lies within 1.96 standard errors of the sample mean. This rule can be used to calculate confidence intervals around estimates and therefore to assess how near

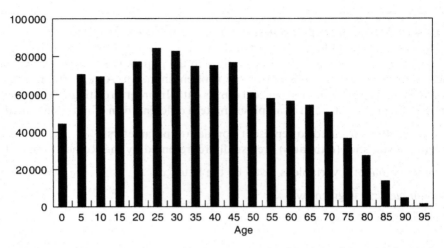

**Fig. 4.2** Histogram of the age distribution of a 2 per cent sample of residents of Great Britain, 1991.
[*Source*: 2 per cent SAR, usual residents, 1991 Census.]

the sample mean is to the population mean. For convenience it is usual to express 95 per cent confidence intervals as ranging from 2 standard errors above or below the mean rather than 1.96.

For example, the average age of residents in the GB Individual SAR is 37.80 years, although this is likely to be a slight underestimate because of top-coding of people over 95 and because the very elderly tend to be under enumerated. A histogram of age taken from the GB Individual SAR (Fig. 4.2) shows that age does not have a normal distribution, but the central limit theorem tells us that, as we have a large sample, the sampling error of the mean of age can be assumed to approximate to a normal distribution. We can use this assumption to calculate confidence intervals around the sample estimate of the mean value of age (see Box 4.3).

In general, there is a straightforward relationship between sample size and the degree of uncertainty about a statistic. The larger the size of the sample and the lower the variability of the item being measured (its standard deviation), the smaller the random difference between the sample and the population. Under the assumption of simple random sampling, the standard error of the mean (SEM) of a variable can be expressed as:

$$\text{SEM} = \frac{s}{\sqrt{n}}$$

where $s$ is the standard deviation of the variable under consideration and $n$ is the sample size. (Note that the standard deviation of the sample is usually taken as an approximation for that of the population.)

From this equation we can see that a fourfold increase in sample size halves the standard error. Thus analysis of small samples will produce much less precise estimates than that of larger samples.

Care should be taken when using very small subsets of microdata. When the sample size is small (less than 30), the normality assumption does not apply and

---

**Box 4.3 *Calculating confidence intervals around the sample estimate of the mean value of age***

The standard deviation of age of GB residents in the 1991 SARs, based on the 2 per cent Individual File is 23.07 and the sample size is 1 080 357.

The standard error of the mean is therefore:

$$\frac{23.07}{\sqrt{1\,080\,357}} = 0.02$$

The standard error of the mean age of the White population is 0.02 and the mean is 38.42.

Assuming a 95 per cent level of confidence, we can calculate confidence intervals (using ±2 standard errors) of

$$38.42 \pm 0.04 = 38.38 - 38.46$$

This represents a range of only 0.08 of a year (30 days), reflecting the very small standard error of the mean which, in turn, is a consequence of the large sample. Assuming that we have a representative simple random sample, we can therefore be 95 per cent confident that the true mean of the age of the resident British population in 1991 lies within the range 38.38 and 38.46 years.

If we now calculate the same estimate for the Bangladeshi population using the 2 per cent SAR, we have a sample size of 3175 and a standard deviation of 17.98. The sample mean is 22.51 years.

The standard error of the mean age of the Bangladeshi population is 0.32, giving 95 per cent confidence intervals of 21.87–23.15 (a range of 467 days).

---

*Student's t-distribution* should be used in constructing confidence intervals rather than the normal distribution. The *t*-distribution is based on the degrees of freedom $(n - 1)$ used in computing the standard deviations and is tabulated for degrees of freedom from 1 to 30. For example, with a sample of 20 cases, there is a 95 per cent probability that the mean lies within 2.09 standard errors of the mean rather than 1.96. The approximate value of 2, however, still serves as a rough guideline for most sample sizes. A more extensive discussion of these issues can be found in most textbooks on sampling and survey methods – Moser and Kalton (1971) and Barnett (1991) provide two excellent examples.

The standard error of the mean is of particular relevance in constructing confidence intervals around the mean of metrical (continuous) variables. However, in SARs, as in most census microdata, most variables are nominal or categorical. For example, marital status, sex and ethnic group are nominal variables whilst highest educational qualification is ordinal – it has an order from low to high but the intervals between categories are not equally spaced. Other variables are ordinal rather than metrical because they have been grouped (e.g. distance travelled to work) or some values have been 'top-coded' to protect confidentiality. For nominal and ordinal variables, statistics such as the mean and standard deviation are not meaningful. However, we may wish to know the number or percentage of people with a particular characteristic or combination of characteristics. Confidence intervals can be constructed around numbers or percentages in a similar fashion to mean values. The following examples are of particular relevance to census microdata.

For a *count* of cases possessing a particular characteristic or characteristics:

$$SE = \sqrt{\frac{c(n-c)}{n}}$$

where $c$ represents the count of cases with the particular characteristics and $n$ the number in the sample.

For a *percentage* of cases possessing a particular characteristic or characteristics:

$$SE(p) = \sqrt{\frac{p(100-p)}{n}}$$

where $p$ is the percentage of cases with a particular characteristic.

Table 4.5 reports the sample cell counts and percentages, their standard errors and confidence intervals for ethnic minority groups in London, using the formulae described above. The relatively small sample errors reflect the large sample size of 123 737.

The standard errors in Table 4.5 are all quite small because even the smallest cell, for the Chinese, does not fall below 1000. However, if we were interested in only the Chinese population ($n = 1087$) and required an estimate of the number who were over 60, the standard error of the estimate would be much larger. The standard error of estimate of Chinese population who are over 60 in London is given by

$$SE = \sqrt{\frac{59 \times (1087 - 59)}{1087}} = 7.47$$

where the cell count, $c$, is 59 and the total number in the sample is 1087.

Therefore the true population size of the Chinese aged over 60 living in London lies in the range 44–74 ($59 \pm 14.5$), at a 95 per cent level of confidence. This range is quite large and alerts us to the need for caution when using small sample numbers. It is also worth pointing out that in five samples out of a 100 we would expect the true population size to be outside this range. At a 99 per cent confidence level (which uses $\pm 3$ standard errors) the range increases to 37–81 ($59 \pm 21.75$). Similarly, if

**Table 4.5** Counts, percentages, standard errors (SEs) and confidence intervals for ethnic minority populations in London (2 per cent Individual SAR: residents only). Sample size = 123 737

|  | Cell count | | Cell percentage | | 95% confidence intervals | |
|---|---|---|---|---|---|---|
|  | $n$ | SE | % | SE | Count | % |
| White | 99 259 | 140 | 80.22 | 0.11 | 98 979–99 539 | 80.0–80.4 |
| Black Caribbean | 5117 | 70 | 4.14 | 0.06 | 4977–5257 | 4.0–4.3 |
| Black African | 2873 | 53 | 2.32 | 0.04 | 2767–2979 | 2.2–2.4 |
| Black Other | 1392 | 37 | 1.12 | 0.03 | 1318–1466 | 1.1–1.2 |
| Indian | 6634 | 79 | 5.36 | 0.06 | 6476–6792 | 5.2–5.5 |
| Pakistani | 1388 | 37 | 1.12 | 0.03 | 1314–1462 | 1.1–1.2 |
| Bangladeshi | 1608 | 40 | 1.30 | 0.03 | 1528–1688 | 1.2–1.4 |
| Chinese | 1087 | 33 | 0.88 | 0.03 | 1021–1153 | 0.8–0.9 |
| Other – Asian | 2173 | 46 | 1.76 | 0.04 | 2081–2265 | 1.7–1.8 |
| Other – Other | 2206 | 47 | 1.78 | 0.04 | 2112–2300 | 1.7–1.9 |

one wants to establish the percentage of elderly Chinese in rented accommodation, confidence intervals can be calculated using the formula for a percentage.

For more complex statistical procedures, statistical software packages normally give standard errors of estimates which can be used in tests of significance (see Ch. 6). However, standard errors derived from estimates based on census microdata should be adjusted due to design effects.

## Design effects

The preceding formulations have been based on the assumption of simple random sampling. However, where the assumptions of simple random sampling do not apply (which is the case for most samples of census microdata as well as survey data), standard errors and the resulting confidence intervals may be misleading. The *design factor* provides an estimate of the effects of the sampling design by comparison with a simple random sample and can be used to adjust the sampling error.

Most sampling designs contain some stratification and also some clustering, both of which affect the sampling error, although in opposite directions. Stratification is used to increase the precision of the sample. In the case of the SARs, stratification was by geographical area, with households listed in blocks of 10 by geographical area and one household selected at random (see Box 4.1 for more details on the sampling strategy used with the 1991 SARs). Clustering is widely used in interview surveys to reduce the cost of travelling. Typically, cluster sampling is geographically based, with a stratified sample drawn within each cluster. The effect of clustering is to increase the size of the sampling error and thus reduce the precision of the estimate. In the 1991 SARs, clustering is particularly apparent amongst individuals in the Household file – explained by the fact that individuals who live together share many other characteristics. The degree of stratification or clustering is measured by the *design effect* of the estimator, where the estimator is either a proportion or a ratio. The square root of this is the *design factor*:

$$\text{Design factor} = \sqrt{\frac{\text{standard deviation of estimator under stratified sampling}}{\text{standard deviation of estimator under random sampling}}}$$

Where the design factor is 1.0, the sampling error is that which would be obtained from a simple random sample. Design factors below 1 occur where stratification gives a sampling error lower than would be expected by simple random sampling. Design factors greater than 1 indicate clustering in the data. Variables may be subject to both simultaneously.

Most sampling procedures are complex and cannot be assumed to approximate to a simple random sample. As explained above, the effect of stratifying the sample by geographical area before sampling reduces sampling error, whilst the inclusion of all household members in the sample causes clustering and increases the sampling error.

Attributes which tend to be shared by all household members (ethnic group, country of birth, migrant status, qualifications, social class) are affected by clustering and have larger than expected sampling errors. Variables which are common across geographical areas (such as housing tenure) may benefit from stratification, producing smaller than expected sampling errors. A variable such as ethnic group may be affected by both stratification and clustering as they are geographically concentrated as well as clustered within household.

Estimated sampling errors for any variable should be modified by multiplying the simple random sampling error by the relevant design factor, which may be estimated separately for each category of each variable. Estimations of design factors in the GB SARs have been calculated using sampling point information (Campbell *et al.*, 1996) and are available for each category of the variables in the Individual and Household SARs.

Although this section is specific to the British SARs, the documentation which accompanies the US PUMS provides extensive information on the sample design, the various sources of error in the data, methods of calculating standard errors and adjustment factors for standard errors.

### Design factors in the GB SAR Household File

In order to illustrate their effect, design factors in the British SARs are discussed in detail below. The sampling for the Household File is subject to geographical stratification and as a result many *household-level variables* (such as housing tenure) have design factors of less than one. Household-level variables relating to particular members of the household (e.g. characteristics of the head of household) are also slightly less than unity.

However, clustering of individual characteristics within households results in sampling errors that are larger than expected for *individual-level variables*. The largest effects are for ethnic group (Table 4.6).

**Table 4.6** Design factors for ethnic group of individuals in the 1991 Household SAR for GB

| Ethnic group | Design factor |
| --- | --- |
| White | 1.84 |
| Black Caribbean | 1.60 |
| Black African | 1.83 |
| Black Other | 1.51 |
| Indian | 1.99 |
| Pakistani | 2.27 |
| Bangladeshi | 2.37 |
| Chinese | 1.87 |
| Other – Asian | 1.83 |
| Other – Other | 1.60 |

These values reflect the fact that people of the same ethnicity tend to live in the same household. The highest values for Pakistanis and Bangladeshis are explained by their larger household sizes. There are also large design factors for country of birth, distance moved (of migrants) and Welsh or Gaelic language indicators. These are all examples of characteristics that tend to be shared by household members. Many other individual-level variables have values near unity.

### Design factors in the GB SAR Individual File

Individual- and household-level variables in the Individual File are less likely to be subject to clustering and may benefit from stratification. However, some clustering effects are apparent because more than one individual may be sampled from the

**Table 4.7** Design factors for ethnic group of individuals in the 1991 Individual SAR for GB, including communal establishments

| Ethnic group | Design factor |
| --- | --- |
| White | 1.01 |
| Black Caribbean | 1.05 |
| Black African | 1.08 |
| Black Other | 1.07 |
| Indian | 1.05 |
| Pakistani | 1.11 |
| Bangladeshi | 1.18 |
| Chinese | 1.19 |
| Other – Asian | 1.11 |
| Other – Other | 1.07 |

same household. However, most design factors are near unity, many being less than 1. The smallest were for household-level variables such as housing tenure and household space type. For individual variables, there are a number of design factors over 1, with the largest, again, for ethnic group, although they are generally much smaller than on the Household File (Table 4.7).

Design factors for a specific geographical area may differ from those for the population as a whole. For example, it has been estimated that whilst the design factor for Black Caribbeans across Great Britain on the Individual SAR file is 1.05, it is 1.12 in Bradford and 1.09 in Birmingham (Campbell *et al.*, 1996). However, design factors are rarely available in such detail and generally the national value should be used.

Where researchers want to use the GB SARs for individual-level analysis and do not need to derive new variables relating to other members of the household (see Ch. 5), they should use the Individual File because its sampling design produces lower design factors. Conversely, those wishing to use household-level variables will find the 1 per cent Household File not only more flexible but probably more representative.

## Using design factors to adjust standard errors

To correct for design effects, the standard errors defined above are simply multiplied by the appropriate design factor. If estimating a count or percentage of persons or households using two or more variables (e.g. Indians in employment), the design factors for the two categories are compared and the largest is used to adjust the sampling error. Once adjusted standard errors of estimates have been calculated, they can be used to calculate confidence intervals, using the formulae given above.

For example, the sample number of Black Caribbeans in London according to Table 4.5 is 5117 with a standard error of 70, giving 95 per cent confidence intervals of 4977–5257. If we apply the design factor of 1.05 for Black Caribbeans, we get a standard error of 73.5 ($1.05 \times 70$). The adjusted confidence interval therefore increases to 4970–5264.

## Combining design factors and population weights

We have seen how inaccuracies can arise from both non-sampling errors and sampling errors. If the user is concerned with making accurate population estimates with reliable confidence intervals, then both population weights and design factors may be used together. However, a further correction factor is required which takes the sample size into account. This is because under-enumeration (and therefore population weights) are correlated with sample size. The following correction factors are suggested (Campbell *et al.*, 1996):

- 1.09 for areas with samples up to 4000
- 1.05 for areas with samples up to 4001–6000
- 1.03 for areas with samples up to 6000+.

In the following section we give a step by step guide to the procedure:

1. Calculate the desired count or percentage; apply the weighting factor to all cases.
2. Calculate the standard error assuming simple random sampling.
3. Adjust the standard error by the appropriate design factor.
4. Multiply the adjusted standard error by the correction factor.

**Worked example: To obtain an estimate of the number of Black Caribbean people in Manchester from the 1991 Individual SAR**

1. Using the SAR data for Manchester, apply the population weights to weight the data back to the Registrar General's mid-year estimate for 1991, by individual year of age and sex.

|  | Unweighted ($c$) | Weighted ($c_w$) |
|---|---|---|
| White | 7135 | 7869 |
| Black Caribbean | 216 | 235 |
| Black African | 80 | 89 |
| Black Other | 93 | 102 |
| Indian | 89 | 103 |
| Pakistani | 302 | 332 |
| Bangladeshi | 53 | 58 |
| Chinese | 71 | 79 |
| Other – Asian | 38 | 41 |
| Other – Other | 96 | 104 |
| Total | 8173 | 9012 |

2. Apply the formula given above to calculate the sampling error of the estimated count assuming simple random sampling:

$$\text{SEM} = \sqrt{\frac{c_w^*(n_w - c_w)}{n}}$$

$$= \sqrt{\frac{235 \times (9012 - 235)}{8173}} = 15.8$$

3. Multiply by the appropriate design factor for Black Caribbeans which has been calculated from the Individual SAR at the national level.

   The appropriate design factor for Black Caribbean is 1.0538

   The corrected sampling error is given by $15.89 \times 1.0538 = 16.7407$

4  Multiply by the relevant correction factor to allow for the fact that under enumeration, and therefore the size of population weights, is correlated with sample size.

The correction factor is $\sqrt{1.03}$ (sample size greater than 6000), giving

$$16.7407 \times \sqrt{1.03} = 16.99$$

Therefore the most accurate estimate of the sample size of Black Caribbeans in Manchester in the 2 per cent SAR will lie within the range $235 \pm 34$ at a 95 per cent level of confidence, or between 201 and 269. This can be converted to a population estimate by multiplying by 50 to allow for the sampling proportion (2/100). The Black Caribbean population for Manchester is therefore estimated as $11\,750 \pm 1700$ or within the range $10\,050-13\,450$ at a 95 per cent level of confidence.

## 4.5 Summary

There are a number of important methodological issues surrounding the analysis of census microdata, reflecting the peculiar nature of such data. These include:

- choosing the appropriate population base for the analysis;
- weighting of data to allow for bias in under-coverage and non-response bias in census data;
- adjusting sampling errors using design factors to allow for the complex sampling procedures used in sample selection;
- simultaneous application of population weights and design factors.

This chapter has also illustrated how, bearing these points in mind, the user can perform exploratory statistical analyses of census microdata, including the calculation of standard errors and confidence intervals of means, counts and percentages. Such analyzes belong to the wider family of inferential statistics which are explored further in Chapter 6 and 7.

# 5

# Variable construction

## 5.1 Introduction

One of the main advantages of census microdata over other types of census output is its flexibility. Researchers are not restricted to using the categorizations given in the microdata, but can group categories in ways appropriate to their own analysis and combine variables to derive new distinctions. In the first part of this chapter, we provide guidance on basic variable construction and then move on to discuss the additional scope which hierarchical data offers, where individuals can be linked within families or households.

## 5.2 Basic variable construction

### 5.2.1 Grouping or recoding into fewer categories

#### Categorical variables

One of the limitations of tabular census data is that they are restricted to categorical variables using predefined groupings. In contrast, microdata usually include much more detail on variables, often more than required for a particular analysis, which users can regroup to suit their requirements. For example, the 1990 US PUMS include considerable detail on place of birth (about 600 categories); language spoken at home (c. 400 categories); race (c. 60 categories); Hispanic origin (c. 80 categories) and ancestry (c. 700 categories). Invariably, categories will need to be grouped before using these variables in an analysis. Some regroupings are provided as additional variables in the dataset; for example, language spoken at home is recoded from about 400 to only six categories: English, Spanish, Other Indo-European, Asian or Pacific Island Language and Other language. Frequently, however, the analysts need to make their own groupings.

The freedom to choose one's own groupings provides enormous advantages – although it also requires a considerable amount of thought in deriving appropriate classifications which best suit the focus of the analysis. For example, the variable on place of birth in the US PUMS may be recoded into a binary variable: USA-born and not USA-born. However, for some research questions (e.g. immigration analyses) greater detail may be required on country or continent of origin. The three-digit code used for place of birth simplifies this as each prefix refers to a particular part of the world. All codes beginning with 0 are for the USA or Puerto Rico, 1 is for Europe, 2 for Asia, 3 for the Americas, 4 for Africa and 5/6/7 for rest of world/at sea/not known. The US PUMS also include a recoded variable for place of birth which combines information on country of birth with citizenship.

Housing tenure – as coded for the 1991 GB SARs – provides another example where regrouping is necessary before any meaningful analysis can be conducted. This is discussed in detail below.

## *Housing tenure: a detailed example*
In the GB SARs, housing tenure has 10 categories: the distribution for households in the 1 per cent file is given below:

| | |
|---|---:|
| 1. Owner-occupied, owned outright | 51 641 |
| 2. Owner-occupied, buying | 91 724 |
| 3. Rented privately, furnished | 7815 |
| 4. Rented privately, unfurnished | 7637 |
| 5. Rented with a job or business | 4214 |
| 6. Rented from a Housing Association or charitable trust | 6809 |
| 7. Rented from a local authority or New Town Development Corporation in England and Wales | 38 385 |
| 8. Rented from a local authority in Scotland | 6761 |
| 9. Rented from a New Town Development Corporation in Scotland | 180 |
| 10. Rented from Scottish Homes | 595 |

To use this variable as it stands could be confusing. First, some of the categories will not be very meaningful to many people – even UK researchers. Second, some categories appear to be similar to others, e.g. categories 7, 8 and 9. Finally, using a variable with 10 categories will produce a rather large table, with some categories containing only a few households, such as renting from a New Town Development Corporation in Scotland. This will lead to large sampling errors and make interpretation difficult.

The first step is to understand what the variable means. This requires locating the census schedules and reading the original question and then understanding how the different categories have been constructed. In the example used here, the question asked in the 1991 Census included a slightly different set of pre-coded answers in England and Wales from those in Scotland, reflecting the fact that housing in Scotland is subject to a different legal and administrative framework than England and Wales. The analyst will need to know that housing associations are charitable bodies which provide social housing, often working closely with local authorities. New Town Development Corporations were set up in the postwar period of the 1950s to manage the growth of new towns, which had been built following slum clearance and wartime bomb damage in older city areas. Scottish Homes is a charitable housing trust specific to Scotland.

Once this information has been established, the analyst can consider how best to group categories; options include:

- a dichotomy between those who own and those who rent (categories 1 and 2, versus all others);
- a trichotomy between those who own, those who rent from a private landlord and those who rent in the 'social housing' sector;
- within the latter group, further distinctions might be introduced between charitable housing (housing associations, trusts, and Scottish Homes) and housing provided by local authorities.

The grouping used will depend upon the analytic focus of the researcher and should clarify, rather than obscure, the distinctions that the researcher wishes to emphasize.

Chapter 4 discussed the various levels of analysis that are typically available in census microdata (e.g. individual, family, household). In the context of this example, it is worth noting that tenure refers to the property in which household members reside. It is therefore a household-level variable and it will usually be appropriate to analyze it at this level although it can also be ascribed to individuals.

Another example, again from the SARs for 1991, is the variable employment status, which has the following codes:

1. Employee full-time
2. Employee part-time
3. Self-employed with employees
4. Self-employed without employees
5. On a government scheme
6. Unemployed
7. Full-time student
8. Permanently sick or disabled
9. Retired
10. Other inactive

This variable may be used to compute a binary variable distinguishing the *economically active* (categories 1–6) from those *economically inactive* (7–10). Alternatively it may be used to distinguish differences in *the relationship to the means of production* for those in work. In this case the distinction between employees (1, 2), the self-employed with employees (3) and the self-employed without employees (4) will be crucial. Decisions will need to be made about how to categorize those on a government scheme. In the mid-1970s, the British Government set up various schemes to provide training to help the unemployed back into work and to provide young people with work experience. Some schemes helped people to set up their own businesses. The schemes have also reduced the number of people who are officially counted as unemployed. The decision on how to categorize those on government schemes will be based on prior knowledge and further exploration guided always by the focus of the analysis. For example, exploratory analysis using the 1991 SARs shows that the majority of those on government schemes are aged 16–18 and involved in youth training programmes; a substantial proportion of these young people did not report an occupation. This may suggest that they should be categorized as unemployed, rather than as employees. It is important to emphasize that there is no one correct answer, but whatever decision is taken should be clearly documented and the rationale behind it given.

## Interval-level variables

Interval-level variables, such as age, income or number of work hours, provide most information when they are retained ungrouped. However, for some purposes it is convenient and helpful to create categories with demarcations which are meaningful in terms of the analysis. For example, one might make different age categorizations according to whether the research focus is labour force activity, family formation or ill-health. In each case one would seek to identify age points which are closely related to aspects of the process under study, e.g. age of entry to the labour market. Similarly one might want to group work hours in relation to break points which are used in labour law and which have implications for benefits and job security. Income variables do not usually have similar break points and therefore it is often helpful to divide an income distribution into deciles or quartiles.

## Checking and documenting derived variables

Whenever new variables are derived, it is essential to check the accuracy of the derivation and to document the process. This can be very simple – by cross-tabulating the target variable (the variable to be derived) with the component variable (the original variable) to check that each category on the component variable has ended up in the correct category of the target variable. Another method of checking, particularly valuable with interval-level variables, is to generate a listing, on a case-by-case basis, of the target and component variables. In most packages this is available (e.g. *summarize cases* in SPSS) and can be done for a sample of cases.

Documenting each variable by adding a variable label and value labels is essential. So, too, is the inclusion of comments to record the reasons for making particular decisions. It should be possible for an outsider to understand how and why variables have been derived and to recreate them in another dataset or using another package. The recommended procedure for creating derived variables is summarized in Box 5.1.

# 5.2.2 Using two or more variables to increase definition

It is often useful to combine information from several variables into a single new variable. From the example of tenure, we can see that in 1991 a majority of British householders either owned their home outright (24 per cent) or were buying it (42 per cent). However, we may wish to make further distinctions between owner-occupiers in terms of their housing assets. For example, we might want to distinguish between those living in a detached house; those in semi-detached or terraced housing, and those in flats or apartments. Using an additional variable in the 1991 SARs, which records a household's type of dwelling space, we can construct the following classification of housing type:

Owner-occupied – detached house
Owner-occupied – other house, e.g. semi-detached or terraced
Owner-occupied – flat/apartment
Rented – private
Rented – public

The SPSS syntax for this is given in Appendix 5.3 (Section A5.3.1).

---

**Box 5.1 Methodological steps in deriving new variables**
1. Define your target variable, i.e. the new variable that you wish to derive.
2. Decide the appropriate level of the variable, e.g. individual or household.
3. Decide the appropriate population, e.g. adults, usual residents.
4. Define the component variable(s) to be used to construct the target variable. At its simplest this may be a single variable; alternatively it may involve a number of variables.
5. Conduct sufficient exploratory analysis to understand the relationship between the component variables.
6. For a complex variable, draw a conceptual mapping of the relationship between the component variables and the target variables.
7. Write the computer syntax for generating the new variable. Include comment commands to explain each step.
8. Add variable and value labels and assign missing values to the new variable.
9. Test that the new variable has been accurately derived:

   • run a frequencies distribution
   • cross-tabulate against the component variables (it is often helpful to include missing values in your cross-tabulation)
   • for a sample of cases, list the value for the component and target variables (list variables in SPSS) so that you can check the derivation for individual cases.

---

## Social classifications based on occupation

All censuses ask basic questions on occupation, employment status and industry. These questions are coded by the census offices into national standardized occupational classifications. If occupational information is held at a sufficiently detailed level, it can be combined with employment status and, sometimes, number of employees in the workplace, to give a range of different class schema. In the UK, the Registrar General's Social Class and Socio-Economic Group are derived by the Census Offices and are available in the SARs, as for other census outputs. However, the basic building blocks of occupation and employment status can be combined in many different ways to give a range of classifications, e.g. Goldthorpe's (1987) classification based upon employment relations; and the International Standard Classification of Occupations (ISCO), which was designed to overcome the problems of comparing occupations that have been coded to distinctive national classification schemes (Section 3.4.3).

## 5.2.3 Constructing more complex classifications or indices

Variables may be combined to create more complex classifications. Simple scales may be constructed to measure the extent to which households lack resources, e.g. the lack of access to a car or the lack of self-contained accommodation. Ideally these should be additive so that lacking both items can be seen as 'worse' than lacking just one or the other (Gordon, 1995). These measures make the assumption (not always justifiable) that the experiences of all members of the household are the same. An example of a very simple measure used with 1991 British data is given below.

## Resources available to the household

This measure is designed to reflect the resources available to the household. The variables chosen can be seen as representing assets that may convert into income (e.g. being in paid work); housing equity; or the opportunities afforded by a car, e.g. access to employment, leisure, schools.

The variables and categorizations are given below:

*Housing*
2    owner-occupied – bought outright (capital asset)
1    owner-occupied – buying on mortgage (may be capital asset)
0    rented housing (no capital asset)

*Central heating – indicating quality/value of housing*
2    heating in all rooms
1    heating in some rooms
0    no heating

*Number of household members in employment*
2+    two or more members in paid work
1    etc.
0

*Number of cars*
2+
1
0

It is assumed that a household which owns its own house (with central heating), has two members in paid work and two cars is in a much better position to generate income and material assets than a household in rented housing, with no-one in paid work and no car. If this scale is summed, it has a range from 0 to 8, with 8 representing households with the most assets. The distribution produced from applying this additive scale is shown in Table 5.1. However, several problems can be identified with this approach. Firstly, it makes the assumption that the scale has equal intervals – that each of the items is equivalent and that the difference between having no car and one car is the same as between having one car and two. Secondly, there are confounding effects caused by household size; one-person households are constrained to have a lower score than households with two or more members. Nonetheless, as Tables 5.1 and 5.2 show, the index gives an approximately normal distribution and shows clear differences between ethnic groups. Whether it is valid to assume that the index is additive can be checked in SPSS using the *reliability* command. In fact, when applied, this test shows that these items are *not* additive. The SPSS commands for deriving this index and using a test for reliability are given in Appendix 5.3 (Section A5.3.2).

**Table 5.1** Index of material resources for households in the 1 per cent SAR for Britain

| Value | Frequency | Percentage |
|-------|-----------|------------|
| .00   | 8747      | 4.1        |
| 1.00  | 10 136    | 4.7        |
| 2.00  | 31 582    | 14.7       |
| 3.00  | 23 151    | 10.8       |
| 4.00  | 31 333    | 14.6       |
| 5.00  | 40 163    | 18.7       |
| 6.00  | 36 299    | 16.9       |
| 7.00  | 28 996    | 13.5       |
| 8.00  | 3962      | 1.8        |
| Total | 214 369   | 100.0      |

**Table 5.2** Mean value and standard deviation on index of material resources for the ethnic group of the head of household, 1991 SARs for Britain

| Value | Mean | SD | Cases |
|-------|------|-----|-------|
| White | 4.3038 | 2.0196 | 206 197 |
| Black Caribbean | 3.6703 | 1.8671 | 2020 |
| Black African | 3.2232 | 1.6934 | 681 |
| Black Other | 3.5913 | 1.9881 | 367 |
| Indian | 5.0638 | 1.7155 | 2148 |
| Pakistani | 4.0229 | 1.8392 | 961 |
| Bangladeshi | 3.2218 | 1.7001 | 266 |
| Chinese | 4.6368 | 1.7287 | 424 |
| Other – Asian | 4.3369 | 1.8723 | 558 |
| Other – Other | 4.0375 | 1.9342 | 747 |
| All ethnic groups | 4.2980 | 2.0161 | 214 369 |

A more sophisticated approach which uses principal components analysis to establish several different dimensions of deprivation is given by Fieldhouse in Chapter 8. It sets out to identify the presence of multiply deprived individuals and draws attention to the fact that using individual-based measures of deprivation to establish the level of need in an area gives somewhat different results from the more widely used measures based on combining aggregate statistics.

## Checking and documenting derived variables

The importance of checking and documenting derived variables cannot be stressed too much. Mistakes are easily made and can have very far-reaching consequences if not identified. Documenting and labelling derived variables, although tedious, are vitally important. One very quickly forgets the details of the decisions made and it is often hard to recreate the rules used as well as the rationale for them.

## 5.2.4 Matching through look-up tables

Using detailed occupational information and employment status, additional variables can be added by means of 'look-up tables' that allocate a value derived from an external data source (Section 2.5.6). For example, a number of continuous variables have been derived to reflect occupational status – the International Socio-Economic Index of Occupational Status (ISEI) (Ganzeboom *et al.*, 1992), the Standard International Occupational Scale (SIOPS) (Ganzeboom and Treiman, 1995), and, specific to Britain, the Cambridge Occupational Scale Score (Prandy, 1990, 1992). These variables are matched onto the individual using a look-up table which holds a previously derived measure of status for each combination of occupation and employment status.

In a similar way, other information can be matched to a set of key variables. One example of this approach is the addition of New Earnings Survey data to the GB SARs. The 1991 GB Census did not ask a question on income. To 'add' information on income, mean hourly earnings, tabulated by occupation, employment status (full- or part-time employment), sex, age and geographical region, were extracted from the New Earnings Survey[1] and mapped to individuals in the SARs using the same key variables to achieve a match (*SARs Newsletter* No. 6, 1995). This aggregate value does not, of course, represent the actual hourly earnings of an individual, but is best understood as an indicator of occupational attainment that provides a mean value for individuals who share the same values on the key variables used. It cannot, therefore, provide information about the effect on occupational attainment of other attributes such as ethnicity. However, this kind of measure of attainment has been widely used in the USA (Roos and Hennessy, 1987; Jasso and Rosenzweig, 1990) and has also been used to provide a British/US comparison of occupational attainment (Ladipo, 1995).

## *Adding area-level classifications*

The addition of area-level classifications to individual or household records provides a further example of adding value to microdata. To retain confidentiality, census microdata are only released for relatively large geographical areas. One way of extending the geographical information available is to attach an area classification – based on a much finer level of geography – to individual or household records. Such a classification will not reveal where individuals live, but provides a description of the kind of area in which they live. In Britain, a ward-level classification has been added to the 1 per cent Household SAR (*SARs Newsletter* No. 9) and an enumeration district[2] classification has been added to the 2 per cent Individual SAR (*SARs Newsletter* No. 12). In both cases, confidentiality criteria have been applied to ensure that there is a minimum number of cases represented within any one category. Examples of descriptors for three different types of area from the GB profile classification added to the Individual SARs for 1991 are:

- elderly, retired home-owners
- Asian, high unemployment, overcrowded, terraced housing
- small, semi-detached council housing.

---

[1] The New Earnings Survey is conducted annually by government. It is based on a sample of employees, using National Insurance number as the sampling frame. It therefore omits the self-employed and employees who are below the 'lower earnings limit' and thus not eligible to pay National Insurance contributions.
[2] In Britain, wards form the lowest level of electoral and administrative geography, typically with about 2000 households. Enumeration districts (EDs) are defined as areas which can be managed by a census enumerator; they typically contain 100–200 households. EDs fit within wards and wards fit within local authority boundaries.

The availability of this area-level information extends the geographical base of the census microdata file and is particularly useful in multilevel modelling applications (Chs 7 and 8). However, the addition of this type of information can only be done by the holders of the original data (the Office for National Statistics in Britain, or the Bureau of the Census in the US) because it requires access to detailed information about place of enumeration.

## 5.3 Hierarchical data

The preceding discussion referred only to data where each case was represented by a single record – usually an individual but sometimes a household. However, where the sampling unit is a household and information is held for all members of the house-hold, there is scope for much creativity in making relationships between members of the same household, e.g. by defining the relative work hours of couples, or the difference in their income levels, or by establishing the relationship between the health of children and the social class or income level of their parents. There is also scope to establish the structure and composition of households and how this varies between ethnic groups (see Ch. 8). This section is concerned with the derivation of variables from hierarchical household data.

The USA, UK and Australian Census Offices release microdata as hierarchical files with the household as the case to which is attached a record for each member of the household (Appendix 2.1). In Australia and the USA, only hierarchical household files are available (with the exception of the non-private household sample from Australia). In the UK two samples are drawn, one of which is a sample of households (which contains records for all household members) and the other a sample of individuals. In Canada there is no household file, although a family file is available which holds information on the partner of each person in a couple. However, this file does not hold linked records for each person in the family. The detailed structure and content of microdata files are discussed in Chapter 2.

It is important to bear in mind the definition of households used in the collection of census data and in census output. Appendix 5.1 (p. 112) gives the definition of house-hold and household head for the countries discussed in this section. In all cases, a notion of common housekeeping is used to define households.

In hierarchical data files from the census there are two 'types' of record (at least): household records and individual person records, as illustrated in Fig. 5.1.

A summary of the definition of families in Britain, the USA and Australia is given in Appendix 5.2 (p. 113). Most households contain only one family; however, for

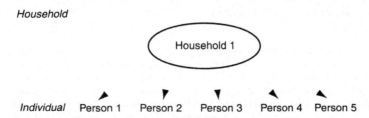

**Fig. 5.1** The relationship between household and individuals.

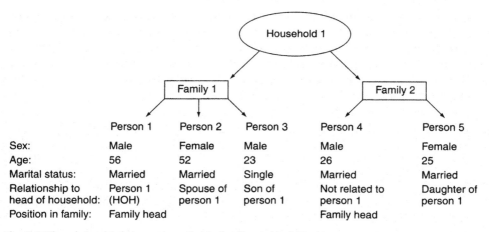

**Fig. 5.2** The relationship between households, families and individuals.

some subgroups of the population it is important to be able to identify a household which contains more than one family. In Britain, a two-family household may comprise a couple living with two children, one of whom is married and also living with a partner. These form two family units: one couple with one never-married child; and one couple (Fig. 5.2). In the SARs, each family is allocated a 'family head' and therefore this household has one household head, but two family heads.

In the USA, subfamilies are identified but are required to be related to the house-holder. In the example given in Fig. 5.2, the married child (person 5) is related to the householder and, with her partner, would form a subfamily. In Britain and Australia, there is no requirement for families to be related to the head of household. In the schedule used in 1991 in Britain, this meant that a family with no relationship to the head of household (first person listed) could not be identified from the question on relationship to head of household, but had, instead, to be identified clerically. In 2001, the UK Census plans to ask about the relationship between each person in the household for households with no more than five residents (HM Government, 1999, Cm 4253). This will increase the accuracy by which families are assigned, although for households containing more than five residents a condensed matrix will be used which may affect coding of large households.

## 5.3.1 Research value of hierarchical data

The ability to understand relationships within the household brings an important dimension to the analysis of census microdata. A number of key areas can be identified:

1. *Household composition.* One of the most valuable uses of hierarchical data is in deriving household classifications. This may be used descriptively to make international comparisons of differences in the composition of households (Wall, 1996); or it may be used to provide comparisons over time – Dale *et al.* (1996) show the increase in one-person households and the decline in two-family

households between 1971 and 1991; or it may allow differences between subgroups of the population to be examined. Some examples of the latter use are given below. The value of census microdata in understanding differences in the household composition of subgroups of the population is amplified by the large sample size. For example, Heath and Dale (1994) showed that, amongst Asian households, young people in partnerships were much more likely to be sharing a household with parents than were young White people.

2. *Ethnic homogeneity*. The 1991 SARs have provided important new information on the ethnic homogeneity of households (Holdsworth and Dale, 1995; Dale and Holdsworth, 1997). This can be expressed as a comparison of the ethnic group of each household member with that of the head of household, or as the ethnic homogeneity of marital or cohabiting unions, or of children and their parents. A comparison of ethnic household composition using the New York PUMS and GB SARs is discussed by Holdsworth in Chapter 8.

3. *Socioeconomic characteristics of household members*. Hierarchical data also allow analysis of the relationship of the socioeconomic characteristics of household members. For example, we can examine the household composition of an unemployed person, and establish the risk of an unemployed young person having other household members who are unemployed. [Payne (1987) provides an analysis using data from the General Household Survey.] Or we can establish the effect of a man's unemployment on his partner's employment status – shown in Table 5.3 for the north-west of England – or the number of dependent children living in households where no-one has paid work. One can also compare work-rich and work-poor households and establish how this relates to stage of life cycle and geographical region.

**Table 5.3** Women's economic activity by that of her husband

|  | Husband | | Total |
|---|---|---|---|
|  | Working | Not working |  |
| Economic position |  |  |  |
| Working | 71.6% | 35.8% | 64.8% |
| Not working | 28.4% | 64.2% | 35.2% |
| Total | 100.0% | 100.0% | 100.0% |

[*Source*: 1 per cent Household SAR, 1991, North-west region]

Section 5.2.2 discussed social classifications based on the occupation and employment status of an individual. Hierarchical data may be used to generate a household measure of social stratification based upon the characteristics of two or more household members or on a reference person within the household. Until the increase in women's labour market participation from the 1960s, it had been commonplace within the sociological literature to attribute to women the class position of their husband. More recently, the justification for this has been hotly debated (Goldthorpe, 1983; Stanworth, 1984; Goldthorpe and Payne, 1986) and alternative measures have been proposed. These have included the 'dominance rule' (Erikson, 1984) which uses the work position of both spouses and is based on an order of dominance where occupations high in this order are assumed to influence the market situation of the family more than lower level occupations. The British Market Research Society (1991) uses the concept of a 'chief income earner' to define the class position of the household. Any of these measures can be applied to census microdata when information is available on each member of the household through a hierarchical file. Measures which relate to couples rather than the household can also be applied to the Canadian Family File which contains individual records for both partners in a family.

The choice of whether to use individual or hierarchical data will therefore depend on the focus of the research. Users of the 1991 GB SARs have a choice between the more detailed geography and larger sample size of the 2 per cent Individual SAR, or the 1 per cent Household SAR with its hierarchical structure and detail on all individuals in the household but less geographical definition. All the Australian and US microdata files are hierarchical, with individuals located within households.

## 5.3.2 Deriving variables

In Britain, Australia and the USA, additional variables describing the composition of households and families within households are available in the files supplied to users. Examples from the SARs include number of residents with limiting long-term illness and number of dependent children in the household; from the Australian Household Sample File, the family type of each family and the relationship between the primary family and other families in the household; and from the USA, employment status of parents and language spoken in the household. However, it is relatively straightforward for users to create their own derived variables.

The way in which hierarchical data is organized may vary with the software used. Some software packages such as SIR are designed for data held in hierarchical form. Thus a single household record is associated with a variable number of individual records. Other software, such as SPSS, requires data to be in a flat or rectangular format with household data repeated for each individual. A household identification variable (SERIALNO in the US PUMS; HNUM in the GB SARs) is attached to each individual record. This can be used as a flag or break variable to identify common membership of the same household. It is clear that software packages based on the hierarchical format avoid the need to repeat household-level information for each household member and therefore data files are smaller. However, because these are used less widely than packages like SPSS, the discussion below is based on the assumption of a flat file. The format of each is shown in Table 5.4.

There are a number of different types of variables that can be derived from hierarchical census microdata for both household and families:

- *Summary variables.* These variables summarize the characteristics of household members. They may either aggregate across all household members, e.g. number of pensioners, number of children, or refer to one member, e.g. age of the oldest person, age of youngest child, chief income earner.
- *Characteristics of reference person.* A reference person is chosen to represent the other members of the family or household, e.g. ethnic group of household head, employment status of family head.

**Table 5.4** Organization of hierarchical data. (a) Hierarchical format; (b) Rectangular format

(a) Hierarchical format

Household 1 (Tenure: owner occupier)

|          | Age | Sex    |
|----------|-----|--------|
| Person 1 | 29  | Male   |
| Person 2 | 27  | Female |

Household 2 (Tenure: renting)

|          | Age | Sex    |
|----------|-----|--------|
| Person 1 | 58  | Male   |
| Person 2 | 55  | Female |
| Person 3 | 20  | Female |
| Person 4 | 23  | Male   |

(b) Rectangular format

|              | Household ID | Household variables | Individual variables | |
|              |              | Tenure              | Age | Sex |
|--------------|--------------|---------------------|-----|-----|
| Household 1  |              |                     |     |     |
| Person 1     | 1            | Owner occupier      | 29  | M   |
| Person 2     | 1            | Owner occupier      | 27  | F   |
| Household 2  |              |                     |     |     |
| Person 1     | 2            | Renting             | 58  | M   |
| Person 2     | 2            | Renting             | 55  | F   |
| Person 3     | 2            | Renting             | 20  | F   |
| Person 4     | 2            | Renting             | 23  | M   |

- *Household classifications.* Classification of households based on the characteristics of individual members (age, sex, marital status) or the number and type of family units. These can also be used to identify family membership.

Summary and reference person variables may be computed for either families or households. Some examples are given below.

## Summary variables

To create a variable summarizing the number of children in a household:

1. count the number of children (NCHILD) with the same household identification number (HNUM) to create a new variable.
2. Save the new variable in a new file along with the household identification variable. This file will hold one record per household.
3. Use this file as a table look-up file to match the variable NCHILD back to the original file.

In SPSS, this procedure is done using the command AGGREGATE to group household members together and compute the new variable which is stored in a separate file (step 1, Fig. 5.3(a)). This new aggregated variable is then MATCHed or linked back

(a) Step 1

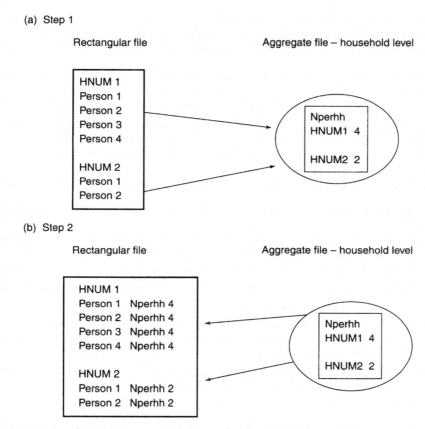

**Fig. 5.3** Derivation of a variable summarizing the number of people in a household.

to the rectangular file, distributing a copy of the household-level variable to each individual (step 2, Fig. 5.3(b)). See Appendix 5.3 (Section A5.3.3) for details of the SPSS commands (p. 115).

## Maximum and minimum variables

The same technique is used to compute maximum and minimum variables. For example, to create a new variable for the age of the youngest dependent child:

1. Derive a new variable for dependent child status using an appropriate algorithm. Compute an age variable which equals the original age for all dependent children and a high number, e.g. 100, for all other household members.
2. Select the minimum value of this new age variable for all cases with the same household identifier. Save to a new file along with household identifier.
3. Match back to the original file.

Appendix 5.3 (Section A5.3.4) gives an example using the 1991 British SARs and SPSS syntax.

## Characteristics of reference person

The value for one particular person in the household can be added to the records of all household members, e.g. ethnic group of household head:

1. Create a new variable for ethnic group which equals the original value for the household head and a set value, e.g. −9, for all other household members.
2. Select the maximum value of the new variable for all cases with the same household identifier, for example if ethnic group is coded 1 to 10, all household heads will have a value in this range, and all other household members will have the value −9. Save this variable to a new file along with the household identifier.
3. Match back to the original file.

Appendix 5.3 (Section A5.3.5) gives an example using the 1991 British SARs and SPSS syntax.

In all the examples above, household variables are linked back to the original dataset. However if the base population is households and not people in households, this is not necessary. For example, for a tabulation of number of cars by ethnic group of household head, it is more appropriate to take households as the base population:

1. Compute the variable for ethnic group of household head as above and save to a new file along with the household identifier. Include in this file the variable for number of cars. As this is the same for each household member, select the first value for each group of cases with the same household identifier.
2. Do not match back to the original file. This new file has one case for each household and can be used to cross-tabulate number of cars by ethnic group of household head.

## Computing family variables

The same techniques can be applied to families, using a family identification variable. In both the GB SARs and the US PUMS, there is a variable for family number (FAMNUM and SUBFAM2, respectively). This variable may be used to modify the household identification variable to identify each family within the household.

For example, in the GB SARs, a family identifier, FAMID, is computed by modifying HNUM:

HNUM +0.1 for individuals in family number 1
HNUM +0.2 for individuals in family number 2
HNUM +0.3 for individuals in family number 3
etc.

This variable may therefore be used in the same way as the household identifier but for the family.

## Family relationships and household classifications

It is also possible to identify each individual's position within a family. For example, in the GB SARs, a variable, FAMTYPE, is included which gives the type of family for each person − based on the definition of a family given in Appendix 5.2 − although, critically, it does not indicate an individual's position in a family. For example, a woman aged 18 living in a one-parent family may be either the parent or a child. However, it is relatively simple to establish position in the family using the following rules:

- couple/couple-parent are the oldest two people in family;
- lone parent is the oldest person in the family;
- a child is any family member not identified as a parent.

It is important to check the accuracy of family relationships, e.g. by checking that there are the same number of male and female partners. In the 1991 GB SARs, a few households were misclassified as they contained children who were older than parents (obviously these were stepchildren, although the British Census did not identify stepchildren). In these households there appeared to be two female or two male partners. These households were identified and recoded using information on marital status. In the 2001 Census, same-sex couples will be included in the coding schema and therefore it will be possible to have different numbers of male and female partners. Moreover, the inclusion of a relationship matrix in the schedule will provide more information on family relationships. In the 1991 GB SARs, it was, in theory, possible to identify concealed cohabiting couples, although in practice only the relationship to the household head was used to reconstruct relationships within households and a large proportion of concealed cohabiting couples were missed in the coding of household relationships.

In the US 1990 PUMS, there is no equivalent to FAMTYPE. The variable RHHFAMTP gives the household/family type which distinguishes households into family and non-family households and, for family households, gives the family type of the householder. For example, a married-couple family household will contain one family of a married couple (including common-law partnerships), one of whom is the householder with or without children and other related/non-related household members. The variable RELAT1 gives relationship to the householder. In households with more than one family (subfamilies in the USA), a variable SUBFAM1 gives the family position of each member of the subfamily. However, there is one important drawback with SUBFAM1, which is that it only identifies married partners in subfamilies and only those who are related to the household head, and hence it is not possible to identify 'concealed' cohabiting couples or couples living with an unrelated household head. This makes comparability of couples in the GB SARs and US PUMS problematic (see Ch. 3) although, as noted above, while the SARs give more information on relationships in complex households, this information is not necessarily complete.

For the US PUMS, family membership can be identified using the following method. For households with one family and for members of the householder's family, the variable RELAT1 will give family membership. For members of subfamilies, use the variable SUBFAM1.

This will identify couples and children; however, it will not necessarily identify parents. To establish whether a couple are also parents, it is necessary to count the number of children in their family, using the technique outlined above for the number of children in the household, modified for families:

1. Create a family identifier variable, FAMID, using SERIALNO and SUBFAM2.

2. Identify all children in the householder's family using the RELAT1 variable, and all children in the subfamilies using SUBFAM1.
3. Count the number of children with the same value of FAMID to create a new variable.
4. Save the new variable in a new file along with FAMID.
5. Use this file as a table look-up file to match the variable on the number of children back to the original file.

All family members will now have a new variable for number of children in the family and it is possible to identify whether each couple has children or not. Information on family membership may then be used to compute a household classification based on the type and number of families in each household.

An SPSS program to compute family membership is given in Appendix 5.3 (Section A5.3.6).

## Ethnic group of children by parent's ethnic group

The strategies outlined above may be used to examine the ethnic group of children by that of their parents. In the 1991 GB Census, 19 per cent of the Black population was coded as Black-Other. This category captured all Black respondents who indicated that they were of neither African nor Caribbean origin. A large proportion of this group were of mixed ethnic origin, whilst others had written in the description 'Black British'. In particular, the majority of Black-Other respondents were children (aged less than 16) and it had been assumed that they were either of mixed ethnicity or 'Black-British' children of Black-Caribbean parents. Analysis of the hierarchical microdata file provided a way of establishing the ethnic group of the parents of Black-Other children (Holdsworth and Dale, 1995). This technique is used to compare the ethnic group of parents for all children (Section 8.4).

The steps in this analysis are set out below:

1. Identify parents and children in family units. For couple-families, the two parents are the oldest two individuals:

   (a) Compute a variable for the age of the oldest person in the household.
   (b) Save to a new file with the *family* identification variable. NB – there will be one case for each family.
   (c) Match the new file from (b) to the original data file.
   (d) Identify the individual in each family whose age equals that of the oldest person.
   (e) Select all other family members, and repeat steps (a)–(d) for the second-oldest person.
   (f) Identify children as all those who are not parents, i.e. not the oldest or second oldest.

   The same technique is used for one-parent families – the oldest person is the parent.

2. Compute the ethnic group of parents:

   (a) Identify the mother from parent and sex variables.

**Table 5.5** Ethnic group of parents of Black-Other children, for children living with both parents

| Ethnic group of father | Ethnic group of mother | | | | | | | | |
|---|---|---|---|---|---|---|---|---|---|
| | White | Black Caribbean | Black African | Black Other | Indian | Chinese | Other – Asian | Other – Other | Total |
| White | 49 | 22 | 10 | 21 | 4 | 2 | 2 | – | 110 |
| Black Caribbean | 117 | 95 | – | 17 | 8 | – | – | – | 237 |
| Black African | 31 | 10 | 4 | 1 | – | – | – | – | 46 |
| Black Other | 40 | 5 | 3 | 80 | 1 | – | 1 | 1 | 131 |
| Indian | 6 | – | – | – | 3 | – | 1 | – | 10 |
| Chinese | 1 | – | – | – | – | – | – | – | 1 |
| Other – Asian | 1 | – | – | 2 | – | – | 13 | – | 16 |
| Other – Other | 2 | – | – | – | – | – | – | – | 2 |
| Total | 247 | 132 | 17 | 121 | 16 | 2 | 17 | 1 | 553 |

(b) Compute a new variable for mother's ethnic group from individual identified in (a).

(c) Repeat for the father's ethnic group.

(d) Save new variables into file with *family* identification variable.

(e) Match new file from (d) to the original data file using the family identification variable.

3. Tabulate parents' ethnic group for Black-Other children:

(a) Select all Black-Other children.

(b) Cross-tabulate variables for mother's ethnic group and father's ethnic group.

SPSS commands for this program are given in Appendix 5.3 (Section A5.3.7). Table 5.5 shows the results from this analysis.

The majority of Black-Other children are of mixed parentage, with 45 per cent having a White mother, although 39 per cent of children have two Black parents divided between Black-Other and Black-Caribbean.

The technique is slightly modified for the US PUMS:

1. Identify couples and children in family units:

(a) For all households with one family and members of the householder's family for multi-family households, identify household heads, partners and children using the RELAT1 and SUBFAM1 variables.

(b) Count the number of children in each household, and save into a file with *family* identification variable.

(c) Match new file from (b) to original data file using family identification variable.

(d) Identify parents using the number of children in family variable.

2. As for GB SARs.

3. As for GB SARs.

## Appendix 5.1: 1990/1 definition of households and household heads

## Great Britain

### Household

- One person living alone; or
- A group of people (who may or may not be related) living, or staying temporarily, at the same address, with common housekeeping.

The definition of common housekeeping was based on either sharing a meal once a day or having a communal living space.

### Household head

The form-filler is asked to give the head or joint head of household as the first person on the form, and this person is then taken as the head of household, provided s/he is aged over 16 and usually resident in the household.

## USA

### Household

A household includes all the persons who occupy a housing unit. A housing unit is a house, an apartment, a mobile home, a group of rooms, or a single room that is occupied (or if vacant, is intended for occupancy) as separate living quarters. Separate living quarters are those in which the occupants live and eat separately from any other persons in the building and which have direct access from the outside of the building or through a common hall.

### Household head (householder)

Person, or one of the persons, in whose name the home is owned, being bought, or rented and who is listed in column 1 of the census questionnaire. If there is no such person in the household, any adult household member 15 years old and over could be designated as the householder.

## Australia

### Household

A group of people who usually reside and eat together. A household is classified as a family household, a group household or a lone person household, or a household containing visitors only, based on relationship information on residents present and temporary absentees. If only children under 15 years of age are present on census night, the household is coded to the category 'not classifiable'. A household resides in a private dwelling (including caravans etc. in caravan parks).

## Household reference person

The family reference person of the primary family (see Appendix 5.2). For group households and households containing only visitors, a household reference person is arbitrarily assigned (ABS, 1991).

## Appendix 5.2: 1990/1 definitions of families

## Great Britain

### Families

Either a couple with or without never-married children or a one-parent family with never-married children. Other people are defined as not being in a family.

### Family head

In the family containing the household head, the same person is taken as the family head; in all other families the family head is either the parent in a one-parent household, or the first couple-member to appear on the census schedule.

(NB this has created an anomaly: in family number one, the family head will be the same as the household head, who is simply the first person aged 16 on the census schedule – there is no control on whether this person is a parent or a child. However, in all families which do not contain the household head, the family head will always be a parent.)

## USA

### Families

A family consists of a householder and one or more other persons living in the same household who are related to the householder by birth, marriage or adoption. All persons in a household who are related to the householder are regarded as members of his or her family. A household can contain only one family for purposes of census tabulations.

### Subfamily

A subfamily is a married couple with or without never-married children under 18 years old, or one parent with one or more never-married children under 18 years old, living in a household and related to, but not including, either the householder or the householder's spouse.

## Australia

### Families

A group of related individuals where at least one person is aged 15 years or over, where related includes related by birth or by marriage or marriage-like (i.e. *de facto*) relationship. Up to three families can be coded in one household: the primary family and up to two 'other' families. If more than three families are found, others are disbanded with their members classified as related family members associated with the primary family or non-family members of the household.

In multi-family households, a family with dependent children is designated as the primary family; if no dependent children are present, the primary family is arbitrarily chosen.

### Family head or reference person
Family relationships are coded based on the relationship to Person 1 on the household schedule. If Person 1 is not the most appropriate (primary) family reference person (e.g. a child), coders assign the reference person based on age, marital status and relationship considerations. For 'other' families in multi-family households, the reference person is assigned by coders. A family reference person must be resident on census night (i.e. not temporarily absent) and aged 15 years or more.

## Appendix 5.3: Examples of SPSS syntax

In the following examples, all SPSS commands are given in UPPER CASE, variable and file names in lower case.

### A5.3.1 SPSS syntax to create a housing type classification based on the 1991 SARs for Great Britain

- Owner-occupied – detached house
- Owner-occupied – other house, e.g. semi-detached or terraced
- Owner-occupied – flat/apartment
- Rented – private
- Rented – public

*Target variable*: house
*Component variables*: tenure, hhsptype

GET FILE 'hholdsar.sav'.

RECODE tenure (1,2 = 1) (3,4,5 = 2) (6,7,8,9,10 = 3) into newten.
VALUE LABELS newten 1 'own-occ' 2 'rent private' 3 'rent public' .

FREQUENCIES newten.

COMPUTE HOUSE = 0.

IF (newten = 1 and hhsptype = 1) house = 1.
IF (newten = 1 and (hhsptype = 2 or hhsptype = 3)) house = 2.
IF (newten = 1 and hhsptype > = 4) house = 3.
IF (newten = 2) house = 4.
IF (newten = 3) house = 5.

VARIABLE LABEL house 'Type of housing'.
VALUE LABELS house 1 'own-occ detached' 2 'own-occ semi-det/terr.' 3 'own-occ flat/apart' 4 'rented private' 5 'rented public'.
FREQUENCIES house.

## A5.3.2 SPSS syntax to compute index of resources and test for scalability

*Target variable*: resindex, index of resources.
*Source variables*: tenure (household tenure), cars (number of cars in household), cenheat (household central heating), dhemp (number of persons in employment resident in household)

GET FILE 'hholdsar.sav'.

* Select household heads from resident population

SELECT IF (relat = 0).
SELECT IF (residsta = 1 or residsta = 2).

* Recode source variables to derive index

RECODE tenure (1 = 1)(2 = 2)(3 THRU 10 = 3) INTO ten.
VALUE LABELS ten 1 'ownout' 2 'ownbuy' 3 'rent'.
RECODE cars (0 = 0) (1 = 1) (2 THRU HI = 2).
RECODE cenheat (1 = 2) (2 = 1) (3 = 0).
VALUE LABELS cenheat 0 'no cenh' 1 'some cenh' 2 'full cenh'.
RECODE dhemp (0 = 0) (1 = 1) (2 THRU HI = 2) INTO nemp.

* Compute index of resources

COMPUTE resindex = SUM (ten + cenheat + cars + nemp).

MEANS TABLES resindex BY ethgroup.
RELIABILITY VARIABLES = ten cenheat cars nemp
   /SCALE (resindex) = ten cenheat cars nemp
   /SUMMARY = MEANS TOTAL.

## A5.3.3 SPSS syntax to derive an aggregate variable and match back to the original file

It may be necessary to sort the dataset by HNUM before matching.
*Target variable*: nchild, number of children aged less than 16 in household
*Source variables*: age, hnum

GET FILE 'hholdsar.sav'.

* Recode age so that 'Under 16' = 1, 'Not under 16' = 0

RECODE age (0 THRU 15 = 1)(16 THRU HI = 0) into childu16.

* Create new file for household variable, number of children under 16

```
AGGREGATE OUTFILE = 'c:\nuchild'
  /BREAK = hnum
  /NCHILD 'No.children under 16 in household'
    = SUM(childu16).
```

* Match new file to original data file

```
MATCH FILES /TABLE = 'c:\nuchild' /FILE = * /BY hnum.
```

```
CROSSTAB TABLES = nchild BY econprim BY sex
  /FORMAT = AVALUE TABLES
  /CELLS = COUNT COLUMN.
```

## A5.3.4 Example of SPSS syntax to derive a minimum household variable from the 1991 GB SARs

*Target variable*: ydepch, age of youngest dependent child in household
*Source variable*: age, econprim, mstatus, hnum

```
GET FILE 'hholdsar.sav'.
```

* Derive a new variable for dependent child status, identifying all children aged 15 and under and those aged 16–18 in full-time education, single and not economically active

```
COMPUTE dchild = 0.
IF age LE 15 dchild = 1.
IF (econprim = 5 AND mstatus = 1 AND econprim = 7 AND (MISSING(econsec)
  OR (econsec > = 5 AND econsec < = 8))) dchild = 1.
```

* Compute newage variable, equals age for dependent children, 100 for all other household members

```
COMPUTE newage = 100.
IF (dchild = 1) newage = age.
```

* Aggregate within households to compute new variable, ydepch, age of youngest dependent child

```
AGGREGATE OUTFILE = 'c:\depchild'
  /BREAK = hnum
  /ydepch = MIN(newage).
```

* Match back to original dataset

```
MATCH FILES /TABLE = 'c:\depchild' /FILE = * /BY hnum.
```

* Frequencies of new variable, ydepch, age of youngest dependent child in household

```
FREQUENCIES ydepch.
```

## A5.3.5 Example of SPSS syntax to derive a variable for characteristic of reference person from the 1991 GB SARs

GET FILE 'hholdsar.sav'.

* Derive a new variable for ethnic group which equals original value for household head and −9 for all other household members.

COMPUTE neweth = −9.
IF relat = 0 neweth = ethgroup.

* Aggregate within households to derive new variable, ethgrphh, ethnic group of household head, use new 'household' file as current file

AGGREGATE OUTFILE = *
  /BREAK = hnum
  /ethgrphh = MAX(neweth)
  /ten = FIRST(tenure).

* Cross-tabulate new variable, ethgrphh, ethnic group of household head, by household tenure

crosstab ethgrphh by ten
  /cells = count column row.

## A5.3.6 SPSS syntax to identify family relationships in US PUMS

* NB it may be necessary to recode variables first to convert them into the correct format

GET FILE 'c:\uspums.sav'.

* Identify all wives and husbands

COMPUTE wife = 0.

DO IF subfam2 = 0 AND rhhfamtp = 1.
IF sex = 1 AND (relat1 = 0 OR relat1 = 1 OR relat1 = 10) wife = 1.
END IF.

COMPUTE husband = 0.

DO IF subfam2 = 0 AND rhhfamtp = 1.
IF sex = 0 AND (relat1 = 0 OR relat1 = 1 OR relat1 = 10) husband = 1.
END IF.

DO IF subfam2 > = 1 .
IF sex = 1 AND subfam1 = 1 wife = 1.
IF sex = 0 AND subfam1 = 1 husband = 1.
END IF.

* Identify all children in families

COMPUTE child = 0.

DO IF subfam2 = 0 AND rhhfamtp > = 1 AND rhhfamtp le 3.
IF relat1 = 2 child = 1.
END IF.

DO IF subfam2 > = 1.
IF subfam1 = 3 child = 1.
END IF.

* Compute family identification variable

COMPUTE famid = serialno.
IF subfam2 = 1 famid = famid + 0.1.
IF subfam2 = 2 famid = famid + 0.2.
IF subfam2 = 3 famid = famid + 0.3.

SORT CASES BY famid.

* Count number of children

AGGREGATE OUTFILE = 'c:\child'
   /BREAK = famid
   /nuchild = SUM(child).

* Match back to original dataset

MATCH FILES TABLE = 'c:\child' /FILE = * /BY famid.

* Identify parents

COMPUTE mother = 0.
COMPUTE father = 0.

DO IF subfam2 = 0 AND sex = 1.
IF (relat1 = 0 or relat1 = 1) AND nuchild > = 1 mother = 1.
END IF.

DO IF subfam2 = 0 AND sex = 0.
IF (relat1 = 0 OR relat1 = 1) AND nuchild > = 1 father = 1.
END IF.

DO IF subfam2 > = 1 AND sex = 1.
IF subfam1 = 1 AND nuchild > = 1 mother = 1.
END IF.

DO IF subfam2 > = 1 AND sex = 0.
IF subfam1 = 1 AND nuchild > = 1 father = 1.
END IF.

FREQUENCIES mother father child.

# A5.3.7 SPSS program to compare children's ethnic group with ethnic group of household head for British SARs

```
GET FILE 'c:\hholdsar.sav'.

SELECT IF residsta = 1 OR residsta = 2.

* Select families

SELECT IF famtype > = 1.

* Compute family identification variable

COMPUTE famid = 0.
IF famnum = 1 famid = hnum + 0.1.
IF famnum = 2 famid = hnum + 0.2.
IF famnum = 3 famid = hnum + 0.3.
IF famnum = 4 famid = hnum + 0.4.

SORT CASES BY famid.

* Compute variable for age of oldest person and save to new file

AGGREGATE OUTFILE = 'c:\oldest'
   /BREAK = famid
   /oldest = MAX(age).

* Match new file to original data file

MATCH FILES TABLE = 'c:\oldest' /FILE = * /BY famid.

* For couples only repeat for second-oldest person

SELECT IF famtype = 2 OR famtype = 3 OR famtype = 4 OR famtype = 5.

SELECT IF age NE oldest.

AGGREGATE OUTFILE = 'c:\second'
   /BREAK = FAMID
   /second = MAX(AGE).

* Return to original file

GET FILE 'hholdsar.sav'.

* Select present and absent residents

SELECT IF residsta = 1 OR residsta = 2.

* Select families

SELECT IF famtype > = 1.

* Compute family identification variable. NB – no need to repeat this if variable is
saved above
```

```
COMPUTE famid = 0.
IF famnum = 1 famid = hnum + 0.1.
IF famnum = 2 famid = hnum + 0.2.
IF famnum = 3 famid = hnum + 0.3.
IF famnum = 4 famid = hnum + 0.4.
```

* Match oldest and second-oldest identification variables

```
SORT CASES BY famid.
```

```
MATCH FILES TABLE = 'c:\oldest' /TABLE = 'c:\second' /FILE = * /BY
   FAMID.
```

* Identify oldest and second-oldest family members as parents

```
COMPUTE parent = 0.
IF (age = oldest OR age = second) parent = 1.
```

* Identify child

```
IF parent = 0 child = 1.
```

* Compute mother's ethnic group

```
COMPUTE moethgrp = −9.
IF (parent = 1 AND sex = 2) moethgrp = ethgroup.
```

* Compute father's ethnic group

```
COMPUTE faethgrp = −9.
IF parent = 1 AND sex = 1 faethgrp = ethgroup.
```

* Save to new file

```
AGGREGATE OUTFILE = 'c:\ethgrp'
   /BREAK = famid
   /mother = MAX(moethgrp)
   /father = MAX(faethgrp).
```

* Match back to original file

```
MATCH FILES TABLE = 'c:\ethgrp' /FILE = * /BY famid.
```

* Select children in couples

```
SELECT IF child = 1 AND (famtype = 2 OR famtype = 3 OR famtype = 5 OR
   famtype = 6).
```

* Cross-tabulate mother's ethnic group by father's ethnic group

```
CROSSTAB mother BY father
   /CELLS = COUNT COLUMN.
```

# 6

# Univariate and bivariate statistics[1]

## 6.1 Introduction: exploratory and inferential statistics

In Chapter 4 we discussed the characteristics of census microdata samples from two perspectives: firstly in terms of the issues inherent in using census data (e.g. choosing population bases, error arising from under-enumeration, imputation and editing) and, secondly, issues inherent in using a sample rather than a population (e.g. sampling error, design effects). In this chapter we move on to preliminary data analysis, again from two perspectives. The first is exploratory data analysis, conducted in order to understand the data structure and check the quality of the data, and the second is basic inferential statistics.

Exploratory data analysis is an essential first step in any analysis of microdata. It provides the means by which the analyst understands how the variables in the database relate to the questions asked in the census; how they have been coded; how non-response has been dealt with; and the distribution of responses on each variable. Unless this preliminary step is conducted, any more sophisticated analysis is likely to be flawed. For analysts new to using samples of microdata it is particularly important to use this exploratory stage to gain an understanding of the way in which choice of population base affects results. In Section 6.2 we discuss standard methods of exploratory data analysis from the perspective of understanding your data. For those who want to pursue this in more detail there are a number of excellent texts available (Marsh, 1988; Tukey, 1977).

In Section 6.3 we discuss inferential statistics for univariate and bivariate relationships. This builds on the discussion, in Section 4.4, of inference from a sample to a population and the assumptions that underlie this: that the data are collected using a simple random sample and are approximately normally distributed. Inferential statistics, which can be extended from univariate analyses to bivariate and multivariate relationships, form the essential underpinnings for the discussion of multivariate methods of analysis and statistical modelling in Chapter 7. For many readers the contents of this section will be very familiar and can be skipped. For those new to the analysis of microdata, we aim to cover enough detail to form a

---

[1] Readers already familiar with basic data analysis are advised to skip this chapter.

bridge to Chapter 7 and provide references to texts for those who wish to follow up topics in more depth. We begin by discussing exploratory statistics.

## 6.2 Exploratory data analysis

Exploratory methods of analysis are an important precursor to more formal statistical tests and methods of analysis designed to allow inferences to the population. Some exploratory methods can also be used in inference. These are particularly important in the early stages of analysis, e.g. in identifying what type of statistical tests may be appropriate or in deciding how to recode or derive variables. In short, it is important to explore the data before launching into sophisticated statistical analyses.

## 6.2.1 Exploring data: univariate analysis

The simplest form of exploratory analysis is to inspect the properties of the distribution of the responses on variables of interest. The notion of a distribution (the pattern of values which a variable takes) was introduced in Chapter 4. A variable has four main properties:

- Level of measurement
  - nominal
  - ordinal
  - interval
- Shape of the distribution
- Outliers
- Spread.

Inspection of these properties will help inform later more sophisticated analyses, including any necessary adjustments or transformations.

### *Level of measurement*
The level at which a variable is measured determines the appropriate kind of statistics that can be used for descriptive and also analytic purposes. With a *nominal* variable, no assumptions are made about the level of the categories or the distance between them. For example, sex has two categories, and marital status may take four or five categories. These have no inherent ordering and therefore measures which assume a metric cannot be applied.

Sometimes, however, a variable may have clearly ordered categories, although the distance between them may not be equal or even readily quantifiable. Social class and educational qualifications are examples of variables that may be measured at an *ordinal level*.

*Interval level* variables, sometimes termed continuous, are so named because they have equal intervals between each, e.g. age or income.

Different summary or descriptive measures and different statistical tests are needed for categorical variables (whether nominal or ordinal) and interval-level variables.

Categorical variables may be displayed as a simple table using a frequencies count or as a bar chart. Both these measures show the distribution of cases across the various categories of the variable and allow the *mode* (the category with the greatest number of cases) to be identified. Interval-level variables may be described using the *mean* or *median* to give a measure of central tendency. A comparison of these two values is helpful in identifying the extent to which the distribution is *skewed* (see below). The distribution may be displayed using a histogram, with a bar for each interval. The size of the bar is proportional to the number or percentage of cases at each point in the specified ranges of values.

## Shape of the distribution

Histograms and bar charts are valuable in showing the shape of the distribution. This may be important when analyzing the data. For example, a *bi-modal* distribution suggests that there may be two different populations which need to be distinguished. An example arises with working hours when the distribution for all workers has two peaks – one at around 38–40 hours and another at around 20 hours, representing full- and part-timers, respectively (Fig. 6.1).

It is often important to know whether a distribution is *symmetrical*. Age is a variable that is asymmetrical – numbers of births are fairly constant each year and deaths do not begin to make a marked impact until the 50s (Fig. 6.2). Household income might be expected to have a fairly symmetrical, normal distribution (Section 4.4); however, histograms invariably show substantial skew towards the upper end of the distribution and a 'floor' at the bottom end where benefit payments prevent income levels falling below a fixed point. Understanding the structure of each variable of relevance is an essential precursor to more sophisticated analysis.

## Outliers

Identifying outliers is another element of the distribution of a variable. Outliers are values which are exceptional in lying at the extreme end of the distribution. With an interval variable such as income, an outlier is a value of income that is distant from other values – either at the top or at the bottom end of the distribution. It may be the result of an error – in either reporting or coding. Although one is not able to go back and examine the census schedule to check on this, it is well worth examining related variables on the record. From this one can establish, for example, whether the person has a job and a level of education that is consistent with the extreme value on income. If the value seems anomalous then it may be appropriate to omit the case from the analysis, or to top-code the variable. Where extreme values appear correct then there may be a need for a transformation (see Box 4.3, p. 143).

By definition, outliers are unusual and are therefore likely to pose a threat to confidentiality. In the SARs for Great Britain, extreme values have been grouped to avoid the risk of identification. For example, all those aged 95 or over have been grouped into a single category and those working over 70 hours have been grouped into two categories: 71–80 and 81 and over. (This grouping, of course, has an effect on the mean and standard deviation of the variables – see below.)

Unusual cases on a categorical variable might also be considered as outliers, e.g. if there was one category of a variable which represented only a tiny proportion of cases. Exploratory analysis of frequency counts would reveal such cases which could

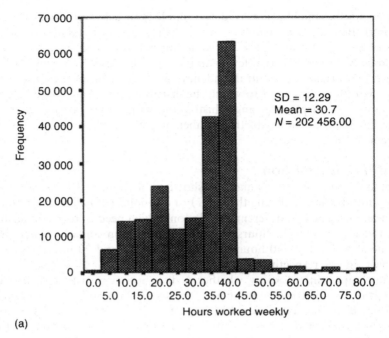

(a)

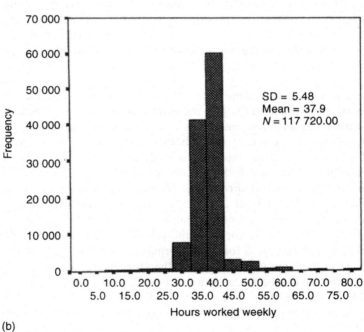

(b)

**Fig. 6.1** Histograms for work hours of women in paid employment (1991 SARs): (a) all women employees; (b) women full-time employees.

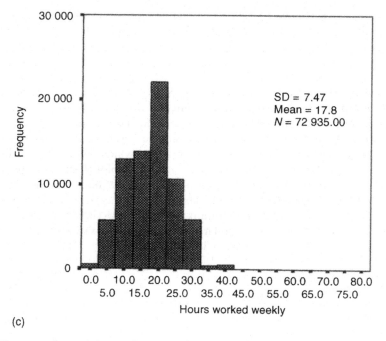

(c)

**Fig. 6.1** Histograms for work hours of women in paid employment (1991 SARs): (c) women part-time employees.

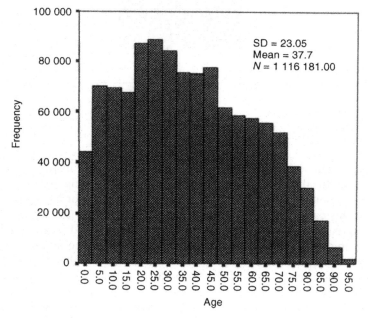

**Fig. 6.2** Histogram of age (SARs 1991).

then be recoded or regrouped to make the variable more amenable to analysis. Sometimes a value occurs which is outside the range of acceptable categories – although the editing procedures that are usually applied to census data make this unlikely.

## Spread

Another aspect of the distribution of a variable is the spread. This simply refers to how much variation there is between cases. For interval data this can be measured using statistics such as the standard deviation or the interquartile range (the mid-spread). The mid-spread represents the distance between the 25th and 75th per centiles (the lower and upper quartiles). This is the range of values which covers the middle 50 per cent of cases. The spread can be represented using a box plot (Fig. 6.3). The box plot (sometimes also called the box-and-whisker plot) shows the median age of the sample. The length of the box represents the interquartile range – i.e. the middle 50 per cent of observations – and maximum and minimum values are also shown. Outliers (between 1.5 and 3 box lengths from the upper and lower edges of the box) may be displayed and, in addition, extreme values – more than 3 box lengths.

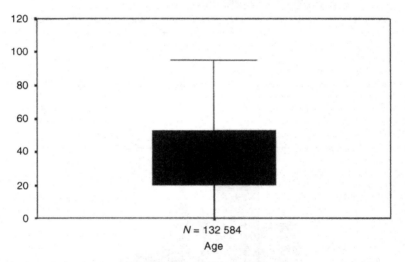

**Fig. 6.3** Box plot of age (London, 1991 Individual SAR).

Spread can also be seen by plotting a histogram where visual inspection allows one to assess whether the distribution is normally distributed (see Fig. 6.2) and the extent to which the distribution is tightly clustered around the mean or has a flatter shaped distribution. This can be formalized using the *standard deviation*, which is a useful measure of spread for interval level variables. The standard deviation is widely used in sampling theory and has special properties under the assumptions of the normal distribution (see Ch. 4). It is proportional to the sum of the squared deviations from the mean and is calculated by the formula:

$$s = \sqrt{\frac{\sum (Y_i - \bar{Y})^2}{N - 1}}$$

In all these measures it is important to remember that where top- or bottom-coding has occurred (exemplified above with age and work hours) then measures of distribution will be affected. In particular, the amount of spread and the number of outliers will be reduced, which, in turn, influences the mean and the standard deviation. However, the median and the mode will not be affected.

These very simple exploratory methods play an important role in checking the coherence and quality of the data and in understanding the nature of the variables to be used. In the next section we move on to consider the value of exploring relationships between two variables.

## 6.2.2 Exploratory bivariate analysis

When planning data analysis, one usually wants to go beyond univariate analyses and to explore relationships between variables and differences between subpopulations. Methods for exploring bivariate relationships can also be used in inference, e.g. by applying confidence intervals or non-parametric tests. These are discussed in more detail below. First we consider those methods most useful in simple data exploration.

### Cross-tabulations
Where variables are categorical (either nominal or ordinal), the most straightforward way of examining a relationship between two variables is by producing a cross-tabulation or a contingency table. One of the major benefits of having census microdata is that it is possible to create tables that combine any combinations of variables. However, as explained in Section 5.2, census variables often contain a large number of categories that make a cross–tabulation very difficult to read. It is therefore helpful to recode or group categories. However, before recoding it is always important to inspect the univariate distribution and to group categories in a way that reflects the theoretical purposes of the analysis.

Interval-level variables may also be grouped, thereby facilitating the use of cross-tabulations to explore relationships. Again, the basis for grouping will reflect the aims of the analysis (Section 5.2).

There are a number of conventions in presenting tabular data that are described below.

### Presenting census microdata as tables
When examining the relationship between two variables, it is often (but not always) assumed that one variable has an effect on the other. In other words, one variable is deemed to be the item of interest (the dependent or response variable) whilst the other is believed to have some influence over the outcome of the dependent variable. For example, we may expect sex to influence economic activity and therefore economic activity would be the dependent variable and sex the independent or explanatory variable. This is the basis of a simple model which, as we see below, can be extended by the addition of a control variable. More complex models are discussed in Chapter 7.

**Table 6.1** Layout for cross-tabulations

(a) **Row percentages**

|  |  | Response variable | Row total |
|---|---|---|---|
| Explanatory variable | → | → | 100 |
|  | → | → | 100 |
|  | → | → | 100 |
|  | Marginal total | Marginal total |  |

(b) **Column percentages**

|  |  | Explanatory variable |  |
|---|---|---|---|
| Response variable | ↓ | ↓ | ↓ |
|  | ↓ | ↓ | ↓ |
| Column total | 100 | 100 | 100 |

When generating a contingency table, the researcher is confronted with a number of choices regarding the format of the table. The following information might be presented

- counts for each cell of the table
- row percentages
- column percentages
- total percentages
- marginal counts and percentages
- expected counts.

It is very hard to read a table that is restricted to cell counts only. Invariably one needs to include some percentages in order to make comparisons – either between the columns or between the rows of the table. It is conventional to construct the percentages so that they sum to 100 within the categories of the explanatory variables. If the explanatory variable is represented by rows the table should look like Table 6.1(a). If the explanatory variable is represented by columns, the table should take the form of Table 6.1(b). Although this table is considered to be the conventional approach in Britain, the decision over which variable to put into rows and which into columns is often dictated by the number of categories: it is usually possible to fit a larger number of categories into rows.

For example, a table of economic activity might be represented as in Table 6.2 (sex being the explanatory variable). This tells us the percentage of men and women in various forms of activity. Activity is represented by rows allowing a greater number of categories to fit on the page. Column percentages are reported, summing to 100 for both men and women. The analysis is limited to people in the age range 16–60.

In the above example, it is clear that gender may influence one's probability of working but the reverse cannot be true. However, in some instances neither variable may be regarded as the dependent variable. Rather, the researcher may simply be interested in the disaggregation of the population into component subpopulations.

**Table 6.2** Economic activity by gender (column percentages)

| Primary economic activity | Men | Women |
|---|---|---|
| Employee full-time | 63.4 | 36.1 |
| Employee part-time | 1.5 | 21.6 |
| Self-employed with employees | 4.1 | 1.3 |
| Self-employed without employees | 8.9 | 2.4 |
| Government scheme | 1.3 | 0.8 |
| Unemployed | 10.1 | 4.7 |
| Student | 5.8 | 5.9 |
| Permanently sick | 3.8 | 3.2 |
| Retired | 0.8 | 1.0 |
| Other inactive | 0.3 | 23.0 |
| All | 100 | 100 |
| $N$ | 316 345 | 3 222 438 |

[*Source*: 2% Individual SAR, 1991, Great Britain]

For example, one may wish to cross-tabulate sex by age group for the population in order to generate a demographic profile or population pyramid (Table 6.3). In these circumstances, it is not necessary to distinguish between dependent and independent variables and a cross-tabulation may simply be presented with cell counts and/or total percentages. Often, however, this type of information would be better presented as a bar chart or histogram.

A number of statistics associated with the table can be calculated, which tell us the likelihood of such a distribution of outcomes having occurred by chance. These are discussed below under inferential statistics.

**Table 6.3** Age by gender (total percentages)

| Age group | Men | Women |
|---|---|---|
| 0–9 | 6.7 | 6.4 |
| 10–19 | 6.4 | 6.2 |
| 20–29 | 7.6 | 7.8 |
| 30–39 | 6.9 | 7.1 |
| 40–49 | 6.8 | 6.8 |
| 50–59 | 5.3 | 5.4 |
| 60–69 | 4.8 | 5.4 |
| 70–79 | 3.0 | 4.2 |
| 80+ | 1.0 | 2.2 |
| Total | 48.5 | 51.5 |
| $N$ | 515 461 | 548 185 |

[*Source*: 2% Individual SAR, 1991, Great Britain]

## Adding control variables

The concept of a two-way contingency table represents the most basic form of model with a response variable and a single explanatory variable. Often, however, one needs a model that allows the relationship between the explanatory variable and the response variable to vary depending on the value of a third variable, called a control variable. The most widely used control variable is probably sex – reflecting the fact that relationships often differ for men and women and, in most analyses, it is important to establish this difference. For example, we may want to explore the impact that

**Table 6.4** Economic activity (a) by dependent children, (b) by dependent children and sex

(a) **Economic activity by dependent children** (row percentages)

|  | In work | Not in work | Total (*N*) |
| --- | --- | --- | --- |
| With dependent child | 68 | 32 | 100 (116 652) |
| Without dependent child | 74 | 26 | 100 (312 995) |

(b) **Economic activity by dependent children and sex** (row percentages)

|  | In work | Not in work | Total (*N*) |
| --- | --- | --- | --- |
| Women |  |  |  |
|    With dependent child | 53 | 47 | 100　(62 727) |
|    Without dependent child | 69 | 31 | 100 (157 826) |
| Men |  |  |  |
|    With dependent child | 85 | 15 | 100　(53 925) |
|    Without dependent child | 78 | 21 | 100 (155 169) |

[*Source*: 2% Individual SAR, 1991, Great Britain]

a dependent child has on economic activity. Does having an extra mouth to feed lead to higher levels of economic activity or do childcare demands make employment less likely? The bivariate relationship between the presence of a dependent child and being in work is explored in Table 6.4(a). Row percentages are given, as the presence of a dependent child is the explanatory variable.

Table 6.4(a) shows a small difference in the percentage of adults who are in work between those with children and those without. However, to understand the relationship fully, we need to bring into our model knowledge of gender roles in Western society and examine the relationship for men and women separately. Table 6.4(b), from the 1991 SARs for Britain, shows the relationship between the presence of dependent children and working, controlling on sex. It becomes clear that the aggregate relationship portrayed in Fig. 6.4(a) concealed very different relationships for men and women. Not only are women less likely to be in work whether or not they have children, but the presence of children has the opposite effect for men as it does for women. Whilst women with children are much less likely to work (only 53% compared with 69%), men with children are more likely to have a job (85% compared with 69%). The bivariate relationship was therefore misleading, particularly for men. Microdata allow for the introduction of any number of control variables appropriate to the analysis, limited only by sample size and, hence, sampling error.

Often our exploratory analyses show us that the distribution of a variable is very different for different age groups and this can indicate the need to control for age. For example, we may expect to find a relationship between reporting a long-term limiting illness and economic activity. However, the effect of this may vary with age. Alternatively, we may wish to compare subgroups of the population based on race or ethnicity that have different age structures (apparent from the example under 'Box plots', below). It is evident, therefore, that a control variable is simply any variable that influences the dependent variable, either directly or through a relationship with some other explanatory variable.

Because microdata allow the researcher the freedom to create tables that combine any combination of variables, it is possible to specify a more complex 'model' by adding control variables. Thus one can build up a model informed by earlier exploratory analysis and also by theory. Subject to the questions having been asked in the census, there is no restriction on the variables that can be included in the model. This therefore allows much greater scope for establishing relationships between variables and building well-informed models than is possible with aggregate census data.

Whilst interval-level variables can be grouped and included in contingency tables, there are other methods of exploratory analysis which may be more useful.

## Box plots

Where one is interested in the relationship between an interval-level variable and a categorical variable, a box plot may be helpful. Whilst box plots can be used to examine univariate distributions (Fig. 6.3), they are probably most useful when making comparisons across subgroups of the population. Figure 6.4 shows a box plot for age for each ethnic group represented in the 1991 SARs for London. The box plot shows the median age of each ethnic group, the length of the box represents the interquartile range and maximum and minimum values; it also indicates any outliers. From this figure we can readily see that there are considerable differences in age structure between ethnic groups – alerting us to the need to control for age on analyses relating to ethnicity. We can also see that, for some groups in particular (Pakistani and Bangladeshi), the age structure is much younger than for other groups such as the Indians.

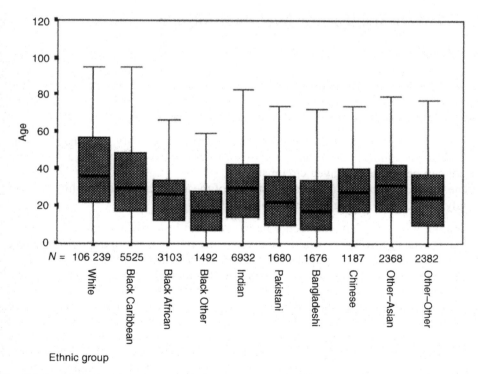

**Fig. 6.4** Box plots of age by ethnic group, London, 1991 Individual SAR

## 6.3 Statistical inference: parametric and non-parametric statistics

In this section we discuss statistics designed to allow us to draw conclusions about the population from which the data were sampled. Inferential statistics differ from exploratory or descriptive statistics only in so far as they draw conclusions about the population.

In most analyses of census microdata, the researcher wishes to generalize from a sample to the population. In Chapter 4 we used parametric statistics to construct confidence intervals (and hence test levels of significance) for means, percentages and proportions for single variables. We used these statistics to infer how well the sample represented the population. Here we extend the use of parametric statistics to analyze bivariate relationships. Many statistical tests are based on certain assumptions about the distribution of the data. The most common is the assumption of 'normality' (Section 4.4), although there are many other distributions upon which statistical tests are based (e.g. the binomial distribution, the $t$-distribution and the Poisson distribution). Such tests are known generically as parametric statistics, after the term 'parameter' which refers to measures describing distributions (e.g. the mean and standard deviation). Non-parametric tests refer to statistical tests that make no assumptions about the underlying distribution of the data. These are also known as 'distribution-free' tests.

### 6.3.1 Tests of significance: hypothesis testing

Statistical inference, whether univariate, bivariate or multivariate, parametric or non-parametric, involves hypothesis testing. For example, when comparing the mean incomes for men and women, we are really testing the hypothesis that there is a difference between male and female incomes. A hypothesis is a statement that can be objectively evaluated from empirical evidence. There are two types of hypothesis:

- The *null hypothesis* (H0) is formulated to test the probability that chance alone could not account for an observation or statistic. It must be written as a statement that can be rejected in given circumstances.
- The *research hypothesis* (H1) is a statement of the anticipated relationship that can only be accepted through rejection of H0.

In Chapter 4 we saw that 'confidence intervals' could be constructed around counts, percentages or means. When calculating confidence intervals, we are, in effect, testing the hypothesis that the true population mean, or a particular count or percentage, lies within a given range (e.g. whether it is significantly different from zero). Alternatively, we can use confidence intervals to deduce whether the mean (or any other statistic) is *significantly different* from that taken from another sample – in other words, is there an overlap in their confidence intervals? This is the same principle as the $z$- or $t$-test described below (Section 6.3.3). If we have taken 95% confidence intervals then we assume the means are 'significantly different' if the confidence intervals do not overlap, and therefore an observed difference is not the chance outcome of simple random sampling. This is termed the 'significance level', which is normally taken as 95 per cent in social research but in some circumstances may be 99 or only 90 per cent. The higher the level of significance, the more certain we are that we have not incorrectly rejected H0.

There are two potential errors in hypothesis testing:

- *type I error*: rejecting H0 when it is, in fact, true;
- *type II error*: failing to reject H0 when it is, in fact, false.

It is widely accepted in scientific method that the burden of proof is on the researcher, and therefore a type II error is normally perceived to be far more acceptable than a type I error. In other words, it is better to accept that something has not been proved than to reject H0 when there remains a considerable degree of uncertainty. For this reason it is usual to set high significance levels (95 or 99 per cent) rather than use the balance of probability (50 per cent).

The results of all statistical procedures can be tested for significance. Most statistical software packages will report a $p$-value for the appropriate statistical test. This represents the probability of the statistic having occurred by chance. If the $p$-value is small (less than 0.05) then this indicates that the statistic is significant at the 95 per cent level (i.e. there is less than a 0.05 probability that the result would have occurred by chance). When using census microdata, sample sizes are usually very large and, because of the relationship between the sample size and the standard error of the mean (SEM),

$$\text{SEM} = \frac{s}{\sqrt{n}}$$

the larger the sample size the more precise the estimate, and thus confidence intervals are smaller (Section 4.4).

In the following sections we review only a small number of the most useful inferential statistics in establishing bivariate relationships. There are, however, a much greater range of measures readily available in many easy-to-use statistical packages. There are two reasons for this limited coverage. The first is that we find very limited use for most of these measures in the analysis of census data. In part this is because one is often dealing with such large sample numbers that almost any differences have statistical significance; the more relevant question is whether observed differences are substantively interesting. In part it is because bivariate analysis is rarely an end in itself but usually a step towards a multivariate modelling approach. The second reason is that there are a great many excellent texts which describe the full range of measures available (e.g. Tukey, 1977; Marsh, 1988; Bryman and Cramer, 1990; de Vaus, 1991; Rose and Sullivan, 1993) and we do not wish to duplicate this material.

Most bivariate analyses using census microdata are based on contingency tables. Therefore we begin by reviewing measures of association for contingency tables with categorical variables. As we cannot make any assumptions about the distribution of these variables, they require non-parametric tests.

## 6.3.2 Non-parametric tests for contingency tables

Our discussion of contingency tables in the context of exploratory data analysis referred to differences in the percentages of men and women who were in paid employment and concluded that men were more likely to be in paid work than women (Table 6.4(a)). From Table 6.4(b) we also concluded that the presence of a dependent child had a negative impact on women's employment but a positive impact on men's. This comparison was made on the basis of the sample and we made no attempt to establish whether the differences that we observed could have arisen by chance. However, what we really wish to claim is that these differences are present in the population. To be able to do so, it is necessary to test the significance of the differences that have been observed. There are a number of non-parametric (distribution-free) tests which are appropriate for categorical variables and available in standard statistical software packages.

## Chi-squared

Pearson's chi-squared test is perhaps the most useful and widely used non-parametric test in analysis of census microdata. It assumes only that the data are from a random sample, and tests the null hypothesis that there is no statistically significant difference between two nominal (categorical) variables. In other words, it indicates whether or not two variables are independent of each other. The chi-squared statistic on which it is based is calculated from the counts in a two-way contingency table (cross-tabulation). The formula compares observed values with those which would be expected if there was complete independence. For example, in Table 6.5, the expected number of men in each category is the number of men that would be found if men and women had identical distributions on economic position:

$$\text{Expected} = \frac{\text{row total} \times \text{column total}}{n}$$

The chi-squared statistic increases with sample size and cannot be interpreted as a measure of association without reference to critical values defined by the chi-squared distribution. Statistical software packages will normally provide the value of chi-squared and the probability of the value arising by chance (a $p$-value or significance level). Chi-squared should not be used for two-by-two tables if there are less than 20 cases or if any cell has an expected count of less than 5. In these circumstances, Fisher's exact test should be used. For larger tables, no expected value should be less than 1 and less than one-fifth of values should be under 5. These conditions

**Table 6.5** Economic activity by sex: City of Edinburgh (residents aged 16+, unweighted)

| Primary economic activity | Men | | Women | |
|---|---|---|---|---|
| | Observed | Expected | Observed | Expected |
| Employee full-time | 1670 | 1312.6 | 1180 | 1537.4 |
| Employee part-time | 65 | 286.9 | 558 | 336.1 |
| Self-employed with employees | 115 | 70.0 | 37 | 82.0 |
| Self-employed without employees | 156 | 94.4 | 49 | 110.6 |
| Government scheme | 32 | 23.9[a] | 20 | 28.1 |
| Unemployed | 238 | 160.7 | 111 | 188.3 |
| Student | 191 | 177.3 | 194 | 207.7 |
| Permanently sick | 132 | 109.1 | 105 | 127.9 |
| Retired | 524 | 591.3 | 760 | 692.7 |
| Other inactive | 23 | 319.6 | 671 | 374.4 |

[a] Minimum expected frequency = 23.9.

are met in Table 6.5, which shows economic activity by sex in the city of Edinburgh:

$$\chi^2 = \sum \frac{(\text{observed} - \text{expected})^2}{\text{expected}}$$

Pearson chi-squared = 1235.921 97

The table has nine degrees of freedom (number of rows $-1$ × number of columns $-1$). The critical value of the test for a 99 per cent significance level is 23.589 (i.e. less than the test value), indicating that there is a significant difference between men and women with respect to economic activity in Edinburgh. Note that the large value of chi-squared in part reflects the large sample size. Most statistical software will report the significance level, but critical values can be easily obtained from published sources (e.g. Neave, 1981).

## *Measures of strength association*

Often one wants to have an indication of the *strength* of an association and, as explained above, the Pearson chi-squared test tells us only whether or not a relationship is statistically significant, and does not indicate the strength of the relationship.

Many statistical software packages will produce measures of association such as the phi coefficient, lambda and Cramer's V, all of which aim to quantify the relationship between two variables. The most useful in analysis of census microdata are those designed for nominal data. These fall into two categories: those based on the chi-squared statistic and those based on the reduction in proportional error. Chi-squared based measures such as phi (for tables with one dimension being two or less), Cramer's V and Pearson's contingency coefficient (for larger tables) all transform the chi-squared to allow for sample size, producing a statistic between 0 and 1. However, these are difficult to interpret as they do not conform to intuitive concepts of association. The values of these tests for Table 6.5 are:

| | |
|---|---|
| Phi | 0.425 36 |
| Cramer's V | 0.425 36 |
| Contingency coefficient | 0.391 42 |

All these test have a *p*-value of less than 0.001, indicating a significant association. Phi and Cramer's V take the same value because one of the dimensions is two or less (sex).

Alternatives to chi-squared measures are those based on a proportional reduction in error, which give a more intuitive indicator of association. Essentially, these measure the extent to which knowledge about one (dependent) variable helps in the prediction of another (independent) variable. Lambda, for example, takes on a value of 0 if the independent variable is of no help in predicting the dependent variable and a value of 1 if the two variables are perfectly matched. The value of lambda varies depending on which variable in a table is considered dependent. Alternatively, a symmetric version of lambda can be calculated, which averages the other two. For example, in Table 6.5 the values of lambda are:

- symmetric                                    0.1180
- with economic activity dependent   0.0000
- with sex dependent                        0.2673

Thus, whilst knowledge of economic activity helps us to predict the sex of an individual, knowledge of sex is of no corresponding benefit in predicting economic activity. This reflects the distribution of cases on the two variables. If the cases are highly skewed on a particular variable (as on economic activity), in order to achieve the highest probability of predicting the correct category, the most frequent outcome (full-time employment) will be predicted regardless of the value on the other variable (sex). In other words, in the example given, full-time employment is the most likely economic activity for men and women. Sex therefore has no effect on the predicted values. In general, measures of association are constructed to measure association in a very specific way and the result will be sensitive to the type of measure used.

### *Rank correlation for relationships between ordinal variables*

Where two variables are ordinal, their relationship can be measured using either Spearman's rho ($\rho$) or Kendall's tau ($\tau$). Coefficients vary between $-1$ and $+1$. These are non-parametric measures of correlation, based on the rank order of cases on an ordinal variable. Neither is commonly used in analysis of microdata because of the lack of ordinal variables. However, they are described in most statistics textbooks. The following section deals more generally with the principle of correlation.

## 6.3.3 Simple parametric tests

In Chapter 4 we saw how we could use the normal distribution to calculate confidence intervals around an estimate for a single variable. Using the same principle as in univariate analyses, we can test whether the mean (or proportion) of a variable is significantly different to the population mean (if it is known), different to a specified value (e.g. zero) or the same for two subpopulations. For example, to examine the relationship between gender and income we can compare the means of the two income groups generated by the binary variable gender (i.e. income for all women and income for all men) using a simple *t*-test or normal distribution *z*-test (see Section 4.4). Although *z* and *t* are calculated in a very similar way, their 'critical values' are different. Essentially, where the calculation of *z* requires knowledge of the standard deviation of the variable in the population, *t* only requires knowledge of the sample standard deviation. Consequently, whilst *z* has a normal distribution, the *t*-statistic has a rather flatter distribution, called the *t*-distribution. The shape of the *t*-distribution means that the smaller the sample size, the larger the difference between two means must be for the *t*-test to prove significant. Effectively, for large samples, *z* and *t* have equivalent values and can be compared against the normal distribution. The value of *t* can be calculated as follows

$$t = \frac{\bar{X} - \mu}{s/\sqrt{n}}$$

where $\mu$ is the population mean, or, alternatively, any hypothetical value against which you wish to compare the sample mean. A situation where the population mean may be known occurs if the sample mean has been calculated from microdata

---

## Box 6.1 Example of simple t-test

To test whether the average earnings of the population in employment in New York is more than \$27 000, we can compare that value with the sample mean from the 1991 PUMS of \$27 170. The sample standard deviation ($S$) has been calculated as \$27 144 and the total sample size is 72 572. The test is one-tailed (we are only interested if it is more than \$27 000)

$$t = \frac{27\,170 - 27\,000}{27\,144/\sqrt{72\,572}} = \frac{170}{100.8} = 1.69$$

The critical value of $t$ for a one-tailed test at a confidence interval of 0.95 (95 per cent) with a large number of cases (where degrees of freedom $= n - 1 = 72\,571$) is 1.6649. If we had wished to set a stricter confidence level of 97.5 per cent, the critical value would be 1.9719. The test value is smaller than this (larger) critical value. We can therefore reject the null hypothesis that income does not exceed \$27 000 dollars with a 95 per cent level of confidence, but not 97.5 per cent.

---

and the population mean is available from the full census data. It is then possible to test whether the sample mean is significantly different from the population mean, and hence whether the sample is representative. Often the population mean will not be known but the researcher wishes to compare the sample mean with a hypothesized value (see Box 6.1). If the value $t$ is greater than the critical value for any given level of significance, the null hypothesis can be rejected. The critical value depends on the level of significance, the sample size (and hence the degrees of freedom) and whether the test is one-tailed (e.g. $x$ bar is greater than $\mu$) or two-tailed ($x$ bar is significantly different from $\mu$).

If, however, we wish to determine whether the means for two independent samples are different then we must calculate $t$ in a slightly different way. For example, it was suggested above that if we wish to examine the relationship between gender and earnings we can compare the means of the two income groups generated by the binary variable gender. If we are fairly certain that the two samples have equal variances, the value of the test statistic for such a comparison is equal to the difference in means divided by the standard error of the difference in means. It is calculated as follows

$$t = \frac{\bar{X}_1 - \bar{X}_2}{s\sqrt{(1/n_1) + (1/n_2)}}$$

where $\bar{X}_1$ is the mean income for men and $\bar{X}_2$ is the mean income for women and $s$ is the pooled sample standard deviation, or the standard deviation of the combined population. The number of degrees of freedom, which determines the critical value of the test, is equal to the combined sample size minus 2.

In many cases we cannot assume that the two samples have equal variances. The assumption of equal variances can be tested using an $F$-test, which is described below. Where the sample variances are known to be unequal, the following formula

---

### Box 6.2 Example of t-test for two independent samples

|       | $N$    | Mean (\$) | Standard deviation (\$) |
|-------|--------|-----------|-------------------------|
| Men   | 38 832 | 33 983    | 31 842                  |
| Women | 33 740 | 19 327    | 17 406                  |

$F = 3298$ (see Section 6.3.4 for calculation of $F$).
Populations do not have equal variance.

$$t = \frac{33\,983 - 19\,237}{\sqrt{(31\,842^2/38\,832) + (17\,406^2/33\,740)}} = 78.7$$

We can therefore reject the null hypothesis that men and women in New York have equal earnings.

This is highly significant as the critical value of $t$ for 99 per cent confidence level is only 2.58. This reflects the fact that not only is there a very large difference between the average earnings of men and women, but, as we have seen on a number of occasions, sample microdata give us a very large sample size.

[*Data source*: 1990 US PUMS with SPSS]

---

applies:

$$t = \frac{\bar{X}_1 - \bar{X}_2}{\sqrt{(s_1^2/n_1) + (s_2^2/n_2)}}$$

where $s_1$ is standard deviation of income for men and $s_2$ is the standard deviation for women. This is illustrated in Box 6.2.

## 6.3.4 Analysis of variance

We have seen that $t$-tests are suitable for comparing pairs of means, where the explanatory variable is dichotomous. However, when analyzing microdata, the researcher often wishes to test whether there is any difference in the mean of a variable between a large number of subpopulations. For example, we may be interested in establishing whether there is a difference in income between different ethnic groups. $t$-tests would be unsuitable for this purpose, as a separate test would be required to compare each possible pair of subgroups. Apart from the inconvenience in calculating a large number of tests, if a large number of pairs of samples were compared, there would be a high probability of a type 1 error (see Section 6.3.1) or, in other words, of achieving some significant test scores by chance. Instead, there is a single test whereby several samples can be compared at once. It is based on a procedure called the analysis of variance (ANOVA) and is called the $F$-test, as used in the Box 6.2 to determine whether two independent samples had equal variances. ANOVA tests the hypothesis that there is a significant difference anywhere amongst the various subpopulations. In order to establish this, analysis of variance compares the variability within groups with that between groups.

Imagine we wanted to compare sample means of income for 10 different samples, representing 10 different ethnic groups. If the variation in income between the 10 groups is relatively small compared with the variability within the groups, it suggests that, in terms of income, the 10 groups can be described as coming from the same population. In other words, there is no difference in income between ethnic groups. If, however, the variation between ethnic groups is large compared with the variability within each group, this would suggest there are differences between groups in their income.

## One-way analysis of variance

One-way analysis of variance is used to compare means where there are three or more subpopulations. The estimate of the between-group variance is compared with an estimate of the within-group variance by dividing the former by the latter. The total amount of variance in the dependent variable (income in our example) can be thought of as comprising that due to ethnic group – the explained variance – and that due to other factors – the unexplained or residual variance. To calculate the $F$-ratio, first we must calculate the *sum of the squares* within and between groups. The sum of the squares was also used in the calculation of the sample standard deviation and variance in Section 4.4.

$$F = \frac{\text{MSB}}{\text{MSW}}$$

where the *mean square between groups* (MSB), or the variance between groups, is obtained by dividing the sum of the squares between groups (SSB) by the number of groups ($K$) minus 1:

$$\text{MSB} = \frac{\text{SSB}}{K - 1}$$

The *sum of the squares between groups* (SSB) is also known as the treatment sum of squares and is calculated as the sum of the squared differences between the group mean and the overall mean weighted by the size of each group:

$$\text{SSB} = \sum_{\text{all groups}} n_i (\bar{x}_i - \bar{x}_{\text{tot}})^2$$

where $n_i$ is the sample size for group $i$ and $\bar{x}_i$ is the sample mean for group $i$. $\bar{x}_{\text{tot}}$ is the mean value from all groups. SSB measures the between-group variability, weighting for the size of each group.

The *mean square within groups* (MSW), or the error mean square, is obtained by dividing the sum of the squares within groups (SSW) by the total number of cases ($N$) minus the number of groups ($K$):

$$\text{MSW} = \frac{\text{SSW}}{N - K}$$

The *sum of the squares within groups* (SSW) is calculated by subtracting the sum of the squares between groups from the total sum of the squares (SS):

$$\text{SSW} = \text{SS} - \text{SSB}$$

---

### Box 6.3 Example of analysis of variance

A 5 per cent subsample of cases from the 1991 US PUMS for New York showed the mean income to be $28 957. However, a breakdown by ethnic groups showed some large differences. An analysis of variance shows that the ratio of the sum of the squared differences between the mean and each individual within groups is approximately 8.5 times as large as the sum of squares between groups:

| | |
|---|---|
| White | $30 050 |
| Black | $22 868 |
| Chinese | $23 340 |
| Other Asian | $31 176 |
| Other race | $18 419 |

Mean square between groups = 6 043 592 152 (five groups = 4 degrees of freedom)
Mean square within groups = 7 077 857 (3667 cases = 3666 degrees of freedom)
Total degrees of freedom = 3670
Critical value at 5 per cent level of confidence = 2.372

$$F = \frac{6\,043\,592\,152}{707\,785\,712} = 8.539$$

Reject null hypothesis that average incomes are equal for different racial groups.

---

where

$$\mathrm{SS} = \sum (X - \bar{X})^2$$

The larger the $F$-ratio, the greater the amount of variation there is between groups. If the value of the $F$-statistic exceeds a critical value then we can reject the null hypothesis that there is no significant difference in means between groups. The critical value of $F$ depends not only on the total number of observations, but also on the number of groups being compared. The $F$-ratio is also used to test the significance of a multiple correlation coefficient ($R$) in a regression model (see Ch. 7).

In the example in Box 6.3, the value of $F$ is easily large enough to be sure (with more than 99 per cent confidence) that there is a difference in the means of income between the racial groups. It is important to note that using the full sample would have given a much larger $F$-value of 185.5. This reflects the fact that a sample of the size often available with census microdata will produce very large values of $F$ where real differences exist. Thus even small differences in the mean will prove to be significant. When using large samples, it is therefore important to distinguish between significance and substantive importance. A $20 dollar difference in mean income may be significant in a very large sample, but is unlikely to be important.

One-way analysis of variance applies in situations where we are simply interested in the difference in means between groups defined by another single variable or factor. It is also commonly known as one-factor analysis of variance. However, often there may be a second factor, or any number of further factors, known to be related to the dependent variable. In other words, we may know that as well as varying by ethnicity, income varies according to gender. In this example, we could perform a two-way analysis of variance, which would allow for differences in gender as well as ethnicity. We would thus be better placed to detect differences between ethnic groups, because (assuming gender had an effect) after allowing for gender, there would be less variance within ethnic groups.

When dealing with census microdata, not only does the researcher have the option of introducing further factors (as one is not constrained to looking at preselected tabulations) but there is often a sufficiently large sample to make this possible. Like all inferential tests, significance in an ANOVA is affected by sample size, not just overall, but within the subgroups analyzed. Most statistical software will perform one-way or two-way ANOVA, giving an $F$-ratio and its significance.

## *Correlation*

As well as being interested in whether there is a 'significant difference' between two populations with respect to a particular variable, in some circumstances we are interested in how one interval-level variable is related to another, without making any assumptions about causality. Ultimately, this allows us to predict the value of one variable if we know the value of another. This principle of prediction is the basis of regression analysis which is explored in more detail in the following chapter. However, correlation is only concerned with measuring the strength of the linear relationship or the association between two variables. One simple way of examining the relationship is to produce a scatter plot. A scatter plot is used to show how two continuous or interval-level variables relate to each other. The scatter plot allows us to visualize the nature and the strength of the relationship between the two variables. For example, if $X$ increased as $Y$ increased, we could say there is a strong positive relationship between $X$ and $Y$ and the points in the scatter plot would form something approximating to a line sloping upwards from left to right (see Fig. 6.5a). If the points were scattered in a more random fashion, we would conclude that the relationship was weak (see Fig. 6.5b). If the relationship was negative, the line would slope from right to left, or in other words as $X$ increases, $Y$ decreases. The strength of relationship is measured by the tendency of points to lie close to an imaginary 'best-fit' line (see Fig. 6.5). This can be quantified and is called the correlation coefficient. The most widely used coefficient of correlation for interval-level data is Pearson's product moment correlation. It is based on fitting the line which minimizes the squared distances between each point on the scatter plot and that line. It is given by

$$r = \frac{S(XY)}{\sqrt{S(XX)S(YY)}}$$

where

$$S(XX) = \sum (X - \bar{X})^2$$

or the sum of the squared deviations of $X$ from the mean value of $X$, and

$$S(YY) = \sum (Y - \bar{Y})^2$$

or the sum of the squared deviations of $Y$ from the mean value of $Y$, and

$$S(XY) = \sum (X - \bar{X})(Y - \bar{Y})$$

or the covariance of $X$ and $Y$.

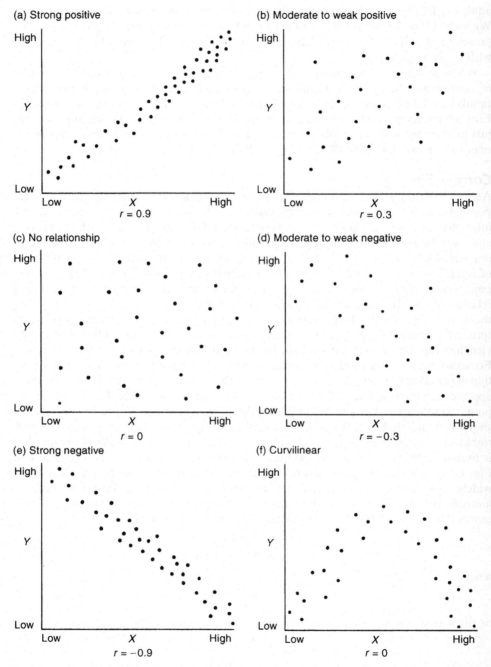

**Fig. 6.5** Scatter plots showing different relationships between two variables. (a) Strong positive; (b) moderate to weak positive; (c) no relationship; (d) moderate to weak negative; (e) strong negative; (f) curvilinear.

The value of $r$ varies between $-1$ and $+1$. A 'perfect' linear relationship between $X$ and $Y$ would exist if all the points fell on a straight line. This would yield a correlation coefficient of 1 for a positive relationship or $-1$ for a negative relationship. A value of 0 would indicate there is no relationship between $X$ and $Y$. The scatter plots in Figure 6.5 show how different values of $r$ represent different relationships between $X$ and $Y$.

The coefficient of correlation only measures the linear relationship. If the relationship was not linear (see Fig. 6.5f) then the correlation coefficient would suggest there was no relationship. It is possible to measure the correlation of two variables which are related in a non-linear fashion by transforming one or both variables (see Box 6.4).

---

## Box 6.4 Transforming the data

The size of $r$, and therefore the strength of the relationship, is measured on the assumption that there is a linear relationship between $X$ and $Y$. If the relationship is not linear then the value of $r$ will not reflect the true strength of the relationship.

For example, the relationship between age and income does not follow a straight line. In order to estimate the strength of the correlation, it is therefore necessary to transform the data so that the relationship is approximately linear. Depending on the nature of the relationship, there are a number of possible transformations (see Marsh, 1988, p. 208). These include:

- square root
- natural log
- log (base 10)
- square
- cube.

Transforming the data can fulfil a number of useful functions. When we are examining or modelling the relationship between two variables, it may be necessary to apply a transformation in order to obtain a linear relationship. It is important to note that by transforming the data we are not attempting to 'fix' the analysis in any way. Rather we are accepting the fact that the relationship between $X$ and $Y$ may not be simple but has a more complex form. How we express the relationship should reflect this. For example, if income rises with age at a steady but decreasing rate, then rather than hypothesizing that income is a function of age, we might instead hypothesize that the log of income is a function of age. This is important when we wish to model relationships.

A second important function of transforming the data is to ensure that the assumptions of any statistical model are not infringed. As we will see in the following chapter, many models assume the dependent variable has a particular form of distribution, very often the normal or Gaussian distribution. By transforming the data it is sometimes possible to obtain a better approximation to the normal distribution. For example, because income is positively skewed – there are many more values near the bottom of the distribution and a small number of people whose incomes are very high – the log of income tends to be roughly normally distributed. This is explored further in the following chapter.

Like all statistics based on sample data, the coefficient of correlation is subject to sampling error, and it is therefore necessary to test whether any association might be the product of chance. In principle, the larger the coefficient, the less likely it is that it has arisen by chance. In other words, if two variables from sample microdata have a correlation of 0.9, it is highly likely that a positive relationship exists between the two variables in the population. In contrast, it is quite likely that a correlation co-efficient of 0.05 might have occurred by chance and that there is no association between the two variables in the population. As for other statistical tests we have described, it is possible to work out the probability of any given value of $r$ having occurred by chance, if the true value in the population was zero. In other words, we are testing the null hypothesis that no relationship exists. The standard error of $r$ can be assumed to have a normal distribution for large samples and can be calculated as

$$s_r = \sqrt{\frac{1 - r^2}{N - 2}}$$

Since the distribution of $r$ is approximately normal and its standard error can be estimated, we can calculate a $t$-ratio and compare it to the $t$-distribution for significance:

$$t = \frac{r\sqrt{N - 2}}{\sqrt{1 - r^2}}$$

If the $t$-value exceeds the critical value at the chosen level of confidence for a two-tailed test (as we are usually interested in whether the coefficient is significantly different from zero) then we can reject the null hypothesis that there is no association between the variables.

Table 6.6 shows the results of a Pearson correlation of age with annual earnings. The correlation of 0.186 is significant at a 99 per cent level. However, if we take the natural logarithmic transformation of earnings (Box 6.4) – and therefore improve the linearity of the relationship between age and earnings – we increase the correlation coefficient to 0.205.

Another important aspect of the relationship between interval-level variables is the gradient or slope of the relationship between $X$ and $Y$. The correlation coefficient tells us only how well the points fit on a best-fit line, or how strong the association is between two interval-level variables, and whether the association is positive or negative. It is important to remember that the correlation coefficient does not tell us anything about causality or about the gradient of the line (the rate of change of $Y$ as $X$ varies). This is known as the regression coefficient and is explored in Chapter 7, where we also consider the causal nature of relationships.

**Table 6.6** Correlation of age with total person earnings, 1989, US PUMS

|  | Earnings in $ | Log of earnings |
| --- | --- | --- |
| Pearson correlation | 0.186 | 0.205 |
| Significance (two-tailed test) | 0.000 | 0.000 |

Correlation significant at the 0.01 level.

## 6.4 Further reading on basic statistics

Bryman, A. and Cramer, D. 1990: *Quantitative data analysis for social scientists*. London: Routledge.

Clegg, F. 1990: *Simple statistics*. Cambridge: Cambridge University Press.

de Vaus, D. A. 1991: *Surveys in social research*. London: UCL Press.

Erickson, B. and Nosanchuk, K. 1979: *Understanding data*. Milton Keynes: Open University Press.

Marsh, C. 1988: *Exploring data: an introduction to data analysis for social science*. Cambridge: Polity Press.

Rose, D. and Sullivan, O. 1993: *Introducing data analysis for social sciences*. Buckingham: OUP.

Rowntree, D. 1991: *Statistics without tears*. London: Penguin.

Tukey, J.W. 1977: *Exploratory data analysis*. London: Addison-Wesley.

# 7

# Multivariate analysis of census microdata

## 7.1 Introduction: statistical modelling

In this chapter we consider the use of statistical modelling techniques in the analysis of census microdata. Statistical models are used in the social sciences to explore and extend understanding of the processes underlying social phenomena such as behaviour or attributes. A fundamental principle of modelling is to identify the characteristic or event of interest, which is termed the *dependent* or *response* variable. The model is used to investigate the impact of other variables, known as *independent* or *explanatory* variables, on the response event or characteristic. The direction of causality is implied, with a change in the explanatory variables causing a change in the response variable. For example, in a model with earnings as the response variable, we may include gender, occupation, qualifications, hours worked, place of employment and ethnicity as explanatory variables. Statistical modelling produces a model equation which quantifies the relationship between the response and explanatory variables. By writing out relationships between variables using equations, we also aim to simplify these relationships, which in turn may help to refine and expand our understanding of complex social processes.

Although this chapter will focus on statistical modelling techniques, it is important to emphasize that such modelling is part of a process that begins with exploratory data techniques discussed in Chapter 6. It is important that the researcher is familiar with the dataset and has an understanding of the relationships between response and explanatory variables. In this way, it is possible to draw up a set of hypotheses regarding the expected relationship between response and explanatory variables and to test these using the modelling techniques. The modelling process may be divided into two parts; first the model is specified, based on theoretical understanding of the phenomena of interest and exploratory analysis of the data. In other words, appropriate explanatory variables should be selected, together with a suitable procedure for estimation. It is then possible to estimate the model from the data. Whilst it is important that a model should always be based on appropriate specification, the process may be iterative. In other words, the model may be re-specified and re-estimated in order to achieve the model which best fits the data. However, when improving the fit of the model by redefining, removing or adding new variables, the theoretical understanding underpinning the model should not be forgotten. A good fit is not a substitute for good theory.

The nature of social science data means that a statistical model will not be able to explain all the variation in the response variable. In any process likely to be of interest here, the amount of individual variation is substantial. Thus it is very unusual for a model to explain more than 50 percent of observed variation, even when it is well specified and all the explanatory variables believed to be important are included. Usually it will be much lower. This highlights an important warning to analysts – that there are many factors that influence outcomes for individuals that cannot be captured by survey data.

In addition, the topic coverage in a census is inevitably more restricted than in a survey designed to answer a specific research hypothesis. For example, in a model to predict limiting long-term illness, the variables available for inclusion may be restricted to housing tenure, social class, marital status and age. Although these will only be able to explain a small amount of the variation in limiting long-term illness (see Ch. 8), the model will, nonetheless, tell us something about the relationship between social characteristics and limiting long-term illness. This reflects the important observation that, rather than attempting to reproduce the complexity of the real world, one of the objectives of modelling is to simplify and make generalizations. The aim of the researcher is to strike a balance between achieving this generalization in a model which is parsimonious, without sacrificing the important complexities of the social phenomena being modelled. Because the full complexity of the real world is never reproduced in the model, all models will be subject to error. The extent to which an individual matches the model's prediction is usually referred to as the 'fit'. The error associated with a particular case is usually termed a 'residual'. We may therefore formalize the relationship between data and explanation, following Plewis (1997):

$$DATA = FIT + RESIDUAL$$

The better the fit of the model, the better the explanation (and the smaller the residual). In statistical modelling, the researcher must aim for a compromise between a model which maximizes the fit and one that is not so complex that interpretation becomes difficult. In other words, while a good fit is important, it should not be achieved at the expense of clarity in the model, as one of the important aims of modelling is to simplify complex social processes. As we shall see later in this chapter, one way of achieving this is to incorporate a structure into our model which reflects that observed in the data.

# 7.1.1 The limitations of cross-sectional data

A census represents a cross-sectional snapshot of the population at a specific point in time. Therefore an explanatory model, which implies a causal ordering, can only be successfully specified where the researcher has a sound theoretical rationale underpinning the choice of response and explanatory variables. We need to be aware that a statistical relationship does not necessarily imply causation. For example, if we wish to explain the incidence of limiting long-term illness, we need a firm theoretical basis for the inclusion of explanatory variables. Although we may find a statistically significant relationship between social class position and health status, this does not

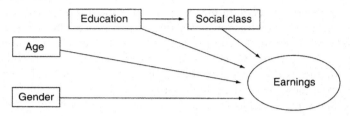

**Fig. 7.1** Diagrammatic model to predict the level of earnings.

tell us the causal direction of that relationship. We rely upon theory, usually based on other empirical studies, to predict that social class will influence health status. In fact, the causal relationship between health and class position has been hotly debated by medical sociologists and it is only with longitudinal data that we can begin to unravel the question of causality (Illsely, 1955, 1986; Power *et al.*, 1990).

It is helpful to draw in diagrammatic form the relationships we expect to observe between explanatory and response variables. The expected relationships will be predicted from prior knowledge or theory but will also have been apparent from earlier exploratory analysis. In Figure 7.1 we have specified a model in which age, gender, education and social class are all expected to have direct effects on earning. However, we also expect that level of education influences social class, and therefore education has both a direct and an indirect effect on earnings. We will see in Section 7.2.4 that this breaks some of the assumptions of linear regression. In later sections we discuss strategies for dealing with this. However, drawing the model in diagrammatic form before embarking on analysis can greatly clarify the expected relationships.

## 7.2 Linear regression

To illustrate the use of linear regression, we use earnings as our response variable. This measure is included in the US and Canadian censuses and is extremely important with respect to people's access to resources and quality of life. We know that earnings levels vary between different groups in society, e.g. between young and middle-aged people, men and women, white collar and blue collar workers and between different ethnic groups. We will therefore take earnings as our response variable and characteristics such as age, sex, ethnicity, educational qualifications as explanatory variables.

However, at this point it is important to be clear about what we mean by earnings. While income may come from a number of different sources – from paid work, from welfare benefits, from capital assets, or from transfers from other household or family members, we usually restrict our definition of earnings to that from paid work. In addition, earnings may be measured at the level of the individual, the family or the household. It is clear that the model used to predict earnings will be different depending on the answers to these questions.

In the example used here, earnings are defined as earnings from all sources (see Box 7.1 for details). We are also taking the individual as the unit of analysis and restricting our attention to adults of working age (i.e. aged 18–65) who were in paid work for at least 40 weeks in 1989 and enumerated as resident in New York in the 1990 PUMS.

## Box 7.1 Data for worked example

*Data source*: US PUMS 1990.

*Population selected*: men and women in employment aged 18–65 inclusive, New York.

*Definition of earnings*: individual's own earnings in 1989 – sum of wage or salary, net farm and non-farm self-employment. Earnings represent the amount of income received regularly before deductions for personal income taxes, social security, union dues, Medicare etc.

The sample is restricted to men and women who worked at least 40 weeks during 1989 and who have non-missing values on earnings.

*Period of time*: Calendar year 1989.

|  |  | Mean | Standard deviation | Min. | Max. |
|---|---|---|---|---|---|
| EARNINGS | Total person earnings, 1989 | 30 157 | 27 742 | 1 | 327 320 |
| Age | Age in years in 1990 | 39.37 | 11.7 | 18 | 65 |
| Gender | Male = 0; female = 1 |  |  | 0 | 1 |
| Race: |  |  |  |  |  |
| Race1 | Black = 1; other = 0 |  |  |  |  |
| Race2 | Chinese = 1; other = 0 |  |  |  |  |
| Race3 | Other Asian = 1; other = 0 |  |  |  |  |
| Race4 | Other race = 1; other = 0 |  |  |  |  |
|  | Number of cases 62 173 |  |  |  |  |

## 7.2.1 Interval-level explanatory variables

In Chapter 6 we looked at the correlation between earnings and age. However the correlation coefficient only measures the strength of the association between two variables and does not describe the relationship between the explanatory and response variables. In order to examine the relationship, a scatter plot may be useful (Fig. 6.5). Figure 7.2 plots a small sample of the data used here. The relationship between the variables can be formalized by the line which 'best fits' the data (Fig. 7.3). The equation which defines this is known as the regression line, and takes the form

$$y_i = b_0 + b_1 x_i + e_i \qquad (7.1)$$

where $y_i$ denotes the response variable and $x_i$ the explanatory variable for the $i$th case; $b_0$ denotes the point at which the line crosses the $y$-axis (the intercept or constant) and $b_1$ is the regression coefficient of variable $x_i$ which represents the gradient or *slope* of the line. The regression coefficient tells us the unit change in earnings ($1) for every increment of age (measured in years). The error term, or residual, is given by $e_i$. We have to make some important assumptions about the error terms; in particular, we assume that they have a mean of zero and unit variance (see below).

For earnings and age, the equation is:

$$\text{EARNINGS} = b_0 + b_1 \times \text{AGE} + \text{RESIDUAL} \qquad (7.2)$$

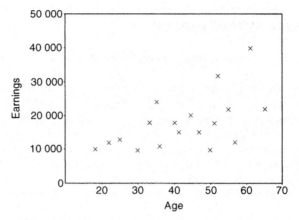

**Fig. 7.2** Scatter plot of relationship between earnings and age.

We can estimate the values for $b_0$ and $b_1$ using a number of techniques, although the most common approach is called least-squares. In this technique the sum of the squares of residuals is minimized or, using statistical notation, we minimize $\sum e_i^2$. The least-squares method also has a number of other desirable statistical properties; in particular, if errors have a normal distribution, least-squares estimates are also maximum likelihood.

The estimated regression line for earnings and age (Table 7.1) is

$$\text{EARNINGS} = 12\,792 + 441 \times \text{AGE} \tag{7.3}$$

The value of $b_0$, the intercept, is \$12 792 and the coefficient for age, $b_1$, is 441. This means that at age 18 earnings are, on average, \$20 730 for members of the sample. This is obtained by multiplying 18 (age) by 441 (the coefficient for age) and adding to the constant (12 792). As age increases by 1 year, on average, earnings increases by a further \$441. We could substitute age into the equation to predict earnings for any given age, e.g. 40:

$$12\,792 + 441 \times 40 = 30\,432 \tag{7.4}$$

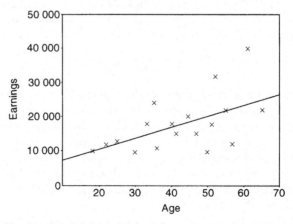

**Fig. 7.3** Regression line draw by minimizing the sum of squares.

**Table 7.1** Results from bivariate regression of earnings on age

| Summary statistics: | $R$ | $R^2$ | Standard error of estimate | | |
|---|---|---|---|---|---|
| | 0.186 | 0.035 | 27 258.3 | | |
| **Coefficients:** | Unstandardized $B$ | Standard error | Standardized $\beta$ | $T(p)$ | Significance |
| Constant | 12 792 | 383.2 | | | |
| Age | 441.0 | 9.32 | 0.186 | 47.3 | 0.000 |

However, we would expect this to be a very poor predictor of earnings as there is only one variable included in the model, and as we will see below, age is only weakly related to earnings.

## 7.2.2 Goodness-of-fit

To address how well this model fits the data, we compute the total sum of squares, which gives the sum of the squared difference between each observed case and the mean:

$$\sum (y_i - \bar{y})^2 \tag{7.5}$$

This total sum of squares is broken down into the regression or explained sum of squares:

$$\sum (\bar{y}_i - \hat{y})^2 \tag{7.6}$$

(or the difference between predicted values of $y$ and the mean) and the residual sum of squares:

$$\sum (y_i - \hat{y})^2 \tag{7.7}$$

A statistic commonly computed from this table is the ratio between the regression or explained sum of squares and total sum of squares, known as $R^2$. The $R^2$ tells us the proportion of variance in the response variable that is explained by the regression equation. In this case (Table 7.1) the $R^2$ of 0.035 indicates that only 3.5 per cent of variance is explained by the regression equation, and we conclude that income is a poor predictor of earnings.

The $R^2$ statistic is summarized as

$$R^2 = \frac{\text{total variance} - \text{residual variance}}{\text{total variance}} \tag{7.8}$$

## 7.2.3 Significance

The $R^2$ tells us about the overall fit of the model. However, we may also want to examine whether each of the variables included in the model is significant, in other words to test whether the true population value of the coefficient is 0. To examine the significance of each coefficient, standard errors are computed (Ch. 4). We usually use the 95 per cent confidence interval to test the null hypothesis that the true value of the coefficient is zero. If the range estimate $\pm 1.96$SE includes zero, then we accept the null hypothesis and conclude that the variable is not significant.

Most statistical packages will calculate a $p$-value which represents the probability of obtaining a given value of $b$ if the true (population) value is zero (i.e. the probability that there is no systematic relationship between $x$ and $y$).

Remember that because sample sizes are usually large when using census microdata, variables entered into the equation will often be found to be significant even where coefficients are small and substantively unimportant. This is because the standard error of the estimate is directly related to the sample size (Ch. 4). Consequently, variables should not necessarily be included simply because of a marginally significant $p$-value.

## 7.2.4 Assumptions of linear regression

Linear regression belongs to the family of inferential statistics known as parametric statistics (Ch. 6). As the name suggests, the validity of parametric statistics is bounded by a number of important assumptions. The inferences that can be drawn from parametric statistics are derived from an assumption about how the data are distributed (see Ch. 4). In the case of linear regression, the fundamental assumption is that the data have a Gaussian or 'normal' distribution (see Box 7.2).

There are various ways to test the validity of the assumptions. The most important are by the examination of the distribution of the response variable (to test for normality); by examining the relationship between the response variable and explanatory variables (to check for linearity); by examining the correlation between explanatory variables; and by inspecting the distribution of residuals and the relationship between residuals. This is normally done using simple scatter plots or histograms as appropriate. More detailed information about testing the assumptions of regression is provided in Fox (1991). Ideally, all the assumptions described in Box 7.2 would be met. The most obvious departure from the assumptions is where the response variable

---

**Box 7.2 Assumptions of linear regression**
- Response variable has a roughly normal distribution
- There is a linear relationship between the explanatory variable and the response variable (which can be transformed if necessary)
- There are no extreme outliers
- No multi-collinearity amongst explanatory variables (there is no strong relationship between them)
- Residuals are independent of each other (not autocorrelated). This is true in a simple random sample
- Variance in residuals is constant and independent of response variable (homoscedasticity)
- Residuals are normally distributed
- Residuals are independent of explanatory variables

is not normally distributed. Most distributions of earnings are highly positively skewed, with a 'floor' at the lower end, particularly in countries with a minimum wage level, and a long tail at the top end of the distribution. However, even where the distribution of a variable in a population is quite clearly not normal, the distribution of repeated sample means or proportions often approaches normality. This is particularly true where the sample size is large. In such cases, the assumption of normality will not lead to misleading results. In other words, linear regression is not particularly sensitive to the assumption of normality. However, if the assumptions of linear regression are badly violated, it may be necessary either to transform the data (see Box 6.4, p. 143) or to use alternative, non-parametric, techniques (see Section 6.3.2).

## 7.2.5 Using residuals to check the assumptions of regression

A residual is the difference between the observed value and the predicted value. The average size of the residuals is used as a basis for measures of goodness of fit, such as $R^2$ and the standard error of the estimate. However, residuals can be directly examined in order to assess whether or not the assumptions of regression are met. The residuals can indicate whether assumptions of linearity have been met and, if not, may indicate the transformations needed. Inspection of the residuals can also indicate whether the assumptions about errors have been met – that error components are independent, have a mean of zero and have the same variance throughout the range of $Y$ values.

The simplest way of checking the normality of residuals is to plot a histogram of their values and to compare it with a normal curve. A second important check is to ensure equality of variance (homoscedasticity). This can be done by plotting the residuals against the predicted values of $Y$. The scatter plot should show no pattern and points should be equally scattered about the line where the value of the residual is zero for all predicted values of $Y$. Most analysis packages will provide scatter plots of the standardized residuals against the predicted $Y$ variable.

## 7.2.6 Transformations to meet assumptions of linear regression

### 1. Are earnings normally distributed?
Figure 7.4 shows a histogram of earnings with a normal distribution imposed. We see that, as expected, earnings have a highly skewed distribution that stretches out at the upper end but is much more condensed at the lower end. A distribution of this shape can be made linear by taking a log transformation (Fig. 7.5).

### 2. Is there a linear relationship between age and earnings?
It is not easy to identify a clear relationship between age and earnings using a scatter plot. However, by plotting mean earnings for each year of age (Fig. 7.6) one can readily see that, by the early 30s, a plateau is reached with little increase in earnings with age and, from the late 50s, a decline.

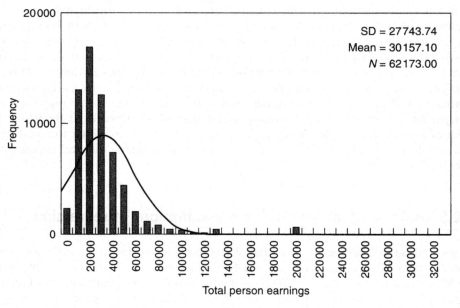

**Fig. 7.4** Distribution of earnings, US PUMS 1990.

These 'bends' can be captured by a polynomial transformation (Box 7.3). Successive powers of the predictor variable – in this example, age – are included in the equation. Thus age squared represents the first turn in the curve at around age 30. To capture the decrease in earnings at older ages, we would also need to add

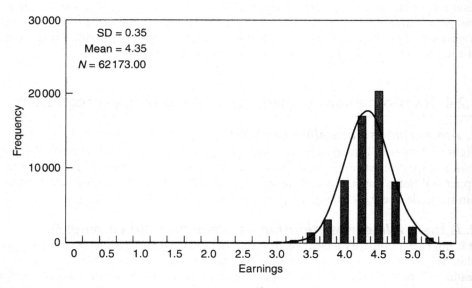

**Fig. 7.5** Distribution of earnings with log transformation, US PUMS 1990.

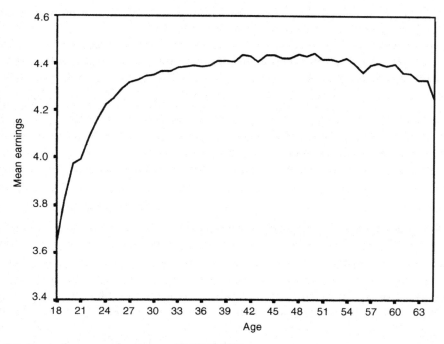

**Fig. 7.6** Graph of mean earnings by age, US PUMS 1990.

---

## Box 7.3 Data transformation

The graph for earnings against age suggests that age is not a good linear predictor of earnings. In this situation we may consider a number of options to transform this relationship.

**Transforming the response variable**

For a response variable such as earnings which is not normally distributed, it is more appropriate to model a function of earnings, e.g. log of earnings.

**Transforming the explanatory variable**

Options are:

- Select an appropriate age range (e.g. all persons aged 18–65);
- Omit extreme earnings values (e.g. all persons earning more than $300 000 per annum);
- Fit other functions of age, e.g. age squared;
- Group age and fit as a categorical variable.

age cubed to the model. By including these extra terms we can transform the relationship between earnings and age to a linear one.

An alternative, which is easier to interpret, is to recode the interval-level variables into categories. Although this reduces the number of degrees of freedom (and hence the significance value of the test), this should not matter with a large sample size, and if it improves the predictive power of the model, it should be considered. This may be particularly useful where there are clear transition points in the relationship between age and earnings. From Figure 7.6 we can see that age 21 appears to be such a transition point. However, categorical variables need to be entered into the model as dummy variables – discussed in the following section.

In the following steps we leave both variables untransformed for ease of interpretation.

## 7.2.7 Categorical explanatory variables

Much of the information collected by censuses (such as social class and marital status) cannot be represented by interval-level variables, but may be either ordinal or categorical. The simplest type of categorical variable is a binary variable, e.g. gender (men, women). We can also reduce more complex categorical variables to simple binary variables; for example, qualifications could be reduced to either 'with qualification' or 'without qualification' and economic activity to 'economically active' or 'not economically active'. If a binary variable is coded 0,1 – e.g. if gender is coded such that 0 represents men and 1 represents women – then this binary variable will denote presence or absence of a characteristic. For example, the variable GENDER indicates male when coded 0 and female when coded 1. This is known as a 'dummy variable' and contrasts the category in question (women) with a base or reference category (men). Categorical variables with more than two categories, such as race or ethnicity, can also be broken down into separate dummy variables (e.g. separate binary variables for White, Black, Indian and Chinese). It is important to remember that when these variables are fitted in a model there should always be one less dummy variable than there are categories. The omitted category is termed a 'reference' category as it is determined by the coding of the other categories. In the example of gender, there is one dummy variable because gender has two categories. If both dummy variables, male and female, were entered into the model, then the assumption of collinearity (independence between the explanatory variables) would be broken, as the category male would be determined by being coded as 'not female'.

To investigate the relationship between gender and earnings, we can estimate a regression equation:

$$\text{EARNINGS} = b_0 + b_1 \times \text{GENDER} \tag{7.9}$$

However, our interpretation is slightly different than that for an interval-level explanatory variable. The variable GENDER can only take two values: 0 or 1. When GENDER equals 0 (i.e. for men) then the term $b_1 \times \text{GENDER}$ also equals 0, and the equation takes the form:

$$\text{EARNINGS} = b_0 \tag{7.10}$$

**Table 7.2** Results from bivariate regression of earnings on gender

| Summary statistics: | $R$ | $R^2$ | Standard error of estimate | | |
|---|---|---|---|---|---|
| | 0.271 | 0.074 | 26 703.8 | | |
| **Coefficients:** | Unstandardized $B$ | Standard error | Standardized $\beta$ | $T(p)$ | Significance |
| Constant | 36 942 | 144.2 | | | |
| Female | −15 132 | 215.3 | −0.271 | −70.3 | 0.000 |

However, when GENDER equals 1 the equation is:

$$\text{EARNINGS} = b_0 + b_1 \tag{7.11}$$

In other words, the coefficient $b_1$ indicates the difference in earnings between men and women. As it is negative we can see that women, on average, have earnings over \$15 000 lower than men (Table 7.2). In other words, the value $b_0$ simply represents the mean earnings for men and $b_1$ represents the differential between men and women. From the $R^2$ shown in Table 7.2 we can also see that more variation in earnings is explained by gender than age – 7.4 per cent by comparison with 3.5 per cent.

We can also fit more complex categorical variables into our model. For example, we may want to look at the impact of ethnicity on earnings, using the variable RACE. This variable is a recoded version of the original census variable with five categories shown below:

0   White
1   Black
2   Chinese
3   Other Asian
4   Other Race (including American Indian)

To include this variable in our regression equation, we need to compute dummy variables as shown in Table 7.3. In this table, the category 'White' is coded 0 in all four dummy variables, which specifies White as the reference category for this particular variable. Black is coded 1 for RACE1 and 0 for all others etc. There are four dummy variables because RACE has five categories.

The regression equation is given by

$$\text{EARNINGS} = b_0 + b_1\text{RACE1} + b_2\text{RACE2} + b_3\text{RACE3} + b_4\text{RACE4} \tag{7.12}$$

**Table 7.3** Constructing dummy variables

| Category in RACE | Value in RACE | Value for each dummy variable | | | |
|---|---|---|---|---|---|
| | | RACE1 | RACE2 | RACE3 | RACE4 |
| White | 0 | 0 | 0 | 0 | 0 |
| Black | 1 | 1 | 0 | 0 | 0 |
| Chinese | 2 | 0 | 1 | 0 | 0 |
| Other Asian | 3 | 0 | 0 | 1 | 0 |
| Other Race | 4 | 0 | 0 | 0 | 1 |

**Table 7.4** Results from regression of earnings on race

| Summary statistics: | $R$ | $R^2$ | Standard error of estimate | | |
|---|---|---|---|---|---|
| | 0.096 | 0.009 | 27 615.8 | | |
| **Coefficients:** | Unstandardized $B$ | Standard error | Standardized $\beta$ | $T(p)$ | Significance |
| Constant | 31 076 | 120.0 | | | |
| RACE1 (Black) | −6470 | 398.7 | −0.065 | −16.2 | 0.000 |
| RACE2 (Chinese) | −5921 | 966.6 | −0.024 | −6.1 | 0.000 |
| RACE3 (Other Asian) | 3054 | 825.6 | 0.015 | 3.7 | 0.000 |
| RACE4 (Other Race) | | 631.3 | −0.069 | −17.2 | 0.000 |

Consider the situation for White respondents: as they are coded 0 for all four dummy variables, the equation is given by

$$EARNINGS = b_0 \tag{7.13}$$

For Black respondents, where RACE1 = 1 the equation is

$$EARNINGS = b_0 + b_1 \tag{7.14}$$

For Chinese respondents the equation is

$$EARNINGS = b_0 + b_2 \tag{7.15}$$

In other words, the coefficient $b_1$ gives the change in earnings for Black respondents *in comparison to* White respondents, and the coefficient $b_2$ gives the change in earnings for Chinese respondents *in comparison to* White respondents.

Table 7.4 shows the results from running this analysis. We can see that the mean earnings for the White group is given by the constant and is $31 076. The earnings for Blacks is given by comparison with that for Whites ($31 076−$6470) and for Chinese ($31 076−$5921). However, for Other Asians it is higher than for Whites ($31 076+$3054). The final category, 'Other Race', is lower again than that for Whites.

## 7.2.8 Multivariate regression

In all the examples so far we have considered a bivariate relationship between the response variable and one explanatory variable. However, we can extend our regression equation to include several explanatory variables:

$$y = b_0 + b_1x_1 + b_2x_2 + b_3x_3 + \cdots + b_nx_n + e \tag{7.16}$$

For example:

$$EARNINGS = b_0 + b_1 GENDER + b_2 RACE1 + b_3 RACE2$$
$$+ b_4 RACE3 + b_5 RACE4 \tag{7.17}$$

When including other explanatory variables in a multivariate analysis, we interpret the coefficient for each explanatory variable as the amount of change in the response for a unit change in the explanatory variable, holding all other variables constant. In other words we are controlling for the effect of other variables. Again, we can see from Table 7.5 the coefficients for each variable. Now, the constant represents the

**Table 7.5** Results from regression of earnings on race and gender

| Summary statistics: | $R$ | $R^2$ | Standard error of estimate | | |
|---|---|---|---|---|---|
| | 0.285 | 0.081 | 26 592 | | |
| **Coefficients:** | Unstandardized $B$ | Standard error | Standardized $\beta$ | $T(p)$ | Significance |
| Constant | 37 701 | 149.5 | | | |
| GENDER (female) | −15 006 | 214.8 | −0.269 | −69.8 | 0.000 |
| RACE1 (Black) | −4952 | 384.6 | −0.050 | −12.9 | 0.000 |
| RACE2 (Chinese) | −5830 | 930.8 | −0.024 | −6.3 | 0.000 |
| RACE3 (Other Asian) | 2245 | 795.1 | 0.011 | 2.8 | 0.000 |
| RACE4 (Other Race | −11 242 | 607.9 | −0.071 | −18.5 | 0.000 |

earnings for the reference category on *both* variables in the equation – white males. The value of $37 097 is about $7000 higher than in the analysis above where it represented the mean for both men and women.

To calculate the mean earnings for White women, we subtract the coefficient representing women from the constant ($37 701−$15 006). Thus women are likely, on average, to have earnings over $15 000 lower than men. To calculate the earnings of Black men, we subtract the coefficient for Black from the constant. However, for Black women we need to subtract the coefficient for women and the coefficient for Black, giving ($37 701−$15 006−$4952).

Although the $R^2$ value has increased to 8 per cent, this is still a very poorly fitting model and we clearly need to add other explanatory variables. For example, we may include age:

$$\text{EARNINGS} = b_0 + b_1\text{AGE} + b_2\text{GENDER} + b_3\text{RACE1} + b_4\text{RACE2}$$
$$+ b_5\text{RACE3} + b_6\text{RACE4} \tag{7.18}$$

The interpretation of the coefficients is similar, with the coefficient for age giving the additional earnings with each year.

These models are examples of main effects models where the impact of each variable is considered in turn. However, we may need to include in our model *interactions* between two variables.

## 7.2.9 Interactions

Linear regression assumes an additive relationship between each of the explanatory variables in the model – that the relationship between the response variable and the explanatory variable is the same for all values of the latter. We have already seen that, with continuous explanatory variables (e.g. age), a transformation can be made to achieve a linear relationship (Section 7.2.4). With categorical variables we may often expect that the relationship with the response variable is different for different categories. For example, we may expect the effect of race on earnings to be different for men and women. By including an interaction term in our model, we can examine this relationship in more detail. Interaction terms are the product of the variables in question. We have four dummy variables for race and one for gender; therefore the interaction terms are

- GENDER × RACE1
- GENDER × RACE2
- GENDER × RACE3
- GENDER × RACE4

As the variables GENDER, RACE1, RACE2, RACE3 and RACE4 can only take the values 1 or 0, their products can only take the values 1 or 0.

The model equation is

$$\text{EARNINGS} = b_0 + b_1\text{AGE} + b_2\text{GENDER} + b_3\text{RACE1} + b_4\text{RACE2}$$
$$+ b_5\text{RACE3} + b_6\text{RACE4} + b_7\text{GENDER} \times \text{RACE1}$$
$$+ b_8\text{GENDER} \times \text{RACE2} + b_9\text{GENDER} \times \text{RACE3}$$
$$+ b_{10}\text{GENDER} \times \text{RACE4} \tag{7.19}$$

To interpret these interaction terms we need to consider combinations of gender and race. For example, White men (the reference category) are coded 0 for GENDER and for all dummy variables representing RACE. Consequently, all interaction terms are also 0, and hence the equation is

$$\text{EARNINGS} = b_0 + b_1\text{AGE} \tag{7.20}$$

For White women, who are coded 1 for GENDER, 0 for all RACE dummy variables and consequently 0 for all interaction terms, the equation is

$$\text{EARNINGS} = b_0 + b_1\text{AGE} + b_2\text{GENDER} \tag{7.21}$$

For Black men, who are coded 0 for GENDER, 1 for the dummy variable RACE1 and 0 for all interaction terms, the equation is

$$\text{EARNINGS} = b_0 + b_1\text{AGE} + b_3\text{RACE1} \tag{7.22}$$

The same applies for all other combinations of GENDER = 0 (i.e. men) and RACE. However, if we turn to consider Black women, the equation is more complicated as this group is coded 1 for both GENDER and RACE1. Thus for Black women the equation is

$$\text{EARNINGS} = b_0 + b_1\text{AGE} + b_2\text{GENDER} + b_3\text{RACE1}$$
$$+ b_7\text{GENDER} \times \text{RACE1} \tag{7.23}$$

The interaction term $b_7\text{GENDER} \times \text{RACE1}$ is included as Black women are coded 1 for both GENDER and RACE1 and are also coded 1 for the interaction term. If we substitute the estimated values for the coefficients into this equation, we find that

$$\text{EARNINGS} = 21\,519 + 424 \times \text{AGE} - 16\,438 \times \text{GENDER}$$
$$- 11\,239 \times \text{RACE1} + 11\,785 \times \text{GENDER} \times \text{RACE1} \tag{7.24}$$

We can interpret the terms in the equation to mean that, allowing for age, in comparison to the reference group White men, Black women suffer a 'penalty' for being a woman and also for being Black. However, the positive interaction term indicates that they earn more than would be expected given the independent effects of ethnicity and gender. A large number of interaction terms can get very complicated to interpret. It is therefore good practice to have a clear theoretical reason for including interaction terms. An alternative to including many interaction terms for a single variable (such as sex) is to fit separate models for different categories or subpopulations. For example, if looking at employment, it may be preferable to fit a separate model for men and women, as different labour market processes are likely to operate for each.

The inclusion of interaction terms always increases the value of $R^2$ (in this case to 0.12), although the increase may not be statistically significant. However, in this example all the interaction terms are significant at a 95 per cent level of confidence.

# 7.2.10 Improving the model

The discussion so far has used variables selected so as to illustrate the steps in model building. In reality, one is likely to build up a model that includes variables with a much more direct impact on earnings than those included above. In this section we discuss models to predict income and earnings from a more substantive aspect, taking into account the availability of data in the US PUMS.

## Measuring income and earnings

An important preliminary step is to find out how income and earnings have been measured. This requires reading the available documentation and, ideally, looking at the 1990 Census schedule to see the wording of the various questions asked. This will indicate which sources have been included and which have been excluded. It will also make clear that figures refer to the previous year, 1989, and that questions were also asked about the number of weeks worked that year and the usual hours worked each week. This means that, for earnings, one can calculate an hourly rate, which takes into account part-time working or a situation where only a few weeks in the year were worked. There will also be decisions to be taken as to whether to include employees and the self-employed or just the former. Earnings information for the self-employed will be much less reliable than for employees.

## Explanatory variables in an analysis to predict earnings

The variables to be included will be informed by one's theoretical perspective. Thus an economist wanting to test a human capital model to explain earnings may include only those variables related to the workers' human capital characteristics (e.g. education, work experience) and omit occupational information. If, however, the question relates to whether job characteristics explain differences in earnings, then the model may include variables which refer to occupation, employment status, industry sector, size of firm and regional location (Joshi and Paci, 1998). A model designed to establish the role played by particular qualifications and the subject of study in relation to earnings will focus on these variables (Dale and Egerton, 1997). Alternatively, one may want to establish the effect on earning of family responsibilities and therefore include number and age of children and presence of a partner (Waldfogel, 1993). The examples given here are drawn from research using British cohort data, but similar models could be used with census data which record earnings. A further influence on earnings may be responsibility for elders, although this is very hard to capture in census data. However, the 1996 Canadian Census asked hours of unpaid care given to elders and the 2001 UK Census plans to ask the number of hours of care provided in addition to care of dependent children. An example of a regression analysis of earnings based on the US PUMS is given in Section 7.2.11.

Earlier exploratory work will have indicated the relationship between earnings and the variables to be used and will give an idea of the likely size and direction of their effect and this will therefore also be used to inform model specification. Drawing the model in diagrammatic form and clarifying the expected direction of influence is of considerable value in ensuring that one has thought through the relationship between the different variables and any necessary interactions.

The next step is to decide how to approach the modelling. The first question to ask is whether the response variable meets the conditions needed for ordinary least-squares (OLS) regression (Box 7.1) or, if not, whether these can be met by using a transformation. OLS is the basic method of regression analysis available in all statistical packages. If this method is chosen, then one needs to decide on how to enter variables into the model. Most packages allow considerable choice. It is often advisable to enter each variable singly as a first stage. This gives basic information on the amount of explanatory value of that variable when considered alone. One can then use this to inform the modelling strategy such that one first enters the variable with the greatest explanatory value and ends with the variable with the least – based on the univariate statistics. The beta coefficient of each variable will alter (usually getting smaller) with each successive variable until one finds that no further variables make a significant difference to the model.

An alternative approach is to base the modelling on theory. For example, when analyzing earnings, as discussed above, one may decide, *a priori*, on the significant variables and enter them in an order that reflects one's theory. Thus variables on educational qualifications and years in school precede occupational information and may therefore be entered first, as a block. Variables relating to current position may then follow, selected for theoretical reasons rather than on the basis of explanatory power. Additional variables may then be added which provide further discrimination – e.g. whether in a male or female dominated occupation, or whether in the public or private sector (see Section 7.2.11). In this case, variables may be retained in the model even if they are not statistically significant, although it is important that they are reported as not being significant.

## Censoring and selection issues

There are a number of circumstances in which ordinary OLS regression will not be appropriate – particularly when data are censored or when there are selection processes that bias the sample for which there is a valid response measure. These are dealt with very comprehensively in Breen (1996). Here, we indicate some of the issues that may arise in the analysis of earnings and income data in samples of census microdata:

1. The response variable is censored – most commonly at the bottom of the distribution. This arises frequently in analyses of income where some individuals do not report any income, although all the explanatory variables may be available for

them. Where this occurs a Tobit model may be considered with maximum likelihood estimation rather than OLS. This is discussed in detail in Breen (1996) who also draws attention to the need to decide on the basis of theory whether or not absence of information should be modelled as a selection process or as censorship. If it is clear that information is not available because of a prior decision by the respondent – e.g. not to take paid work – then it is that decision-making process that should be modelled, rather than treating the data as censored. This is discussed in the following section.

2. Sample selection is typically modelled as a two-stage process in the analysis of earnings. We find a higher proportion of working age men in employment than women – however, we are only able to observe earnings for those who are in work. Differences vary between countries and over time and are likely to be greater in Britain than in the USA. In Britain in 1998, 90 per cent of men aged 21–60 were in employment compared with 70 per cent of women. A substantial proportion of women in Britain leave the labour market upon the birth of their first child and remain out of employment for a number of years. However, women who are not in employment differ significantly from those who are – particularly in having lower educational qualifications and, in general, less earning capacity. This means that, when making comparisons between the earnings of men and women, there is a sample-selection bias. This is widely recognized by economists who typically use a two-stage process whereby the selection process into employment is modelled first and, if significant bias is found, the results are used to correct the earnings equation.

There are unmeasured factors that influence both the decision to take paid work or not and also how much one earns if working. These factors are often described as 'motivation' and lead to correlated errors. Breen (1996) argues that this should not be seen as misspecification that would be corrected if 'motivation' were adequately measured, but instead should be seen as intrinsic to the process and modelled as such.

A two-step estimation can be readily implemented in STATA using Heckman's procedure (Heckman, 1979). In the earnings example, a participation probit is first modelled using all women in the sample, not just those in work, to obtain the probability of working (probit models are discussed further in Section 7.3). The probit result is then used to compute a variable – the 'inverse Mills ratio', usually denoted as lambda, $\lambda$ – to reflect the probability that a woman with the same characteristics as the sample member is working. A linear regression can then be used with the subsample of working women and $\lambda$ included along with the other explanatory variables (Breen, 1996). This gives parameter estimates corrected for selection bias.

There are mixed opinions on the extent to which there should be overlap between the variables in the two equations. Generally it is best not to use all the variables in the selection equation in the outcome equation as this makes assumptions about the form of the probit analysis that may not always be met. Breen (1996) provides a full discussion of the more technical issues involved. In the example in Section 7.2.11, less than half the variables in the participation probit are also used in the earnings equation. An extremely helpful discussion of the issues around sample selection can be found in Berk (1983).

Appendix 7.1 gives the STATA commands for running a Heckman's two-stage selection procedure to model participation as a multinomial process, e.g. as full-time, part-time and not working.

# 7.2.11 An example of an analysis of earnings based on the US PUMS

Tienda *et al.* (1987) provide a valuable example of the use of 1970 and 1980 US PUMS to explain the 1980 earnings gap between men and women. They capitalize on the benefits to be gained from census data by including in their analysis a measure of industrial restructuring between 1970 and 1980, and also changes in occupational segregation by sex between these two time points. The latter measures were computed for men and women separately from a set of industry by occupation matrices for 1970 and 1980 and are described in detail in an appendix to the paper.

The variables included in their regression analysis are shown in Table 7.6. The modelling strategy adopted by Tienda *et al.* (1987) was to run a succession of models which regressed annual earnings for each sex separately, first using only human capital variables, race/ethnicity variables and the control variables, including $\lambda$ to correct for selection bias, and then introducing, in succession, variables for occupational mix and industrial shift, proportion female in 1980, and feminization rate.

Tienda *et al.* suggest that sample selection bias may occur for both men and women, particularly when a sample is restricted to those in employment in a particular year.

**Table 7.6** Variables used in regression analysis of earnings by Tienda *et al.* (1987)

| Variable | Definition |
|---|---|
| *Dependent variable*: earnings | Wage and salary EARNINGS in 1979 |
| *Human capital measures* | |
| Age | Age in years |
| Age$^2$ | Square of age |
| Education | Highest grade of school completed |
| High school graduate | 1 if high school graduate; 0 if not |
| College graduate | 1 if college graduate; 0 if not |
| English proficiency | 1 if proficient in English; 0 if not |
| *Race/ethnicity* | |
| Hispanic | 1 if Hispanic origin; 0 if not |
| Black | 1 if black; 0 if not |
| *Structural change indices* | |
| Occupational mix | Intra-industry occupational mix, 1970–80 |
| Industrial shift | Industry shift, 1979–80 |
| Proportion female | The proportion of female workers in each of 335 job cells in 1980 |
| Feminization rate | (% female 1980 – % female 1970)/ % female 1970 |
| *Controls* | |
| Married | 1 if married; 0 if not |
| Household head | 1 if household head; 0 if not |
| Foreign born | 1 if foreign born; 0 if not |
| Employed full-time | 1 if employed full-time; 0 if not |
| $\lambda$ | Inverse of Mill's ratio, predicted from reduced-form probit equation for being in wage sample |

[*Source*: Tienda *et al.* (1987), Table 2]

**Table 7.7** Variables used in participation probit by Tienda *et al.* (1987)

| Variable used in probit analysis |
| --- |
| Age in years |
| Age$^2$ |
| Education: highest grade of school completed |
| Currently married |
| Household head |
| Children less than 6 in the household |
| Children aged 6–11 in the household |
| Children aged 12–18 in the household |
| Hispanic |
| Black |
| Foreign born |
| Disabled |
| Non-wage EARNINGS |
| Area wage |
| Area unemployment rate |

[*Source*: Adapted from Tienda *et al.* (1987), Table A-1]

They therefore extracted their value $\lambda$ for both men and women using a probit analysis with 15 explanatory variables, five of which were common to both equations. The variables used in the probit analysis to predict the probability of being in the wage equation are shown in Table 7.7.

Although not discussed by Tienda *et al.*, the results of the participation probit show interesting differences between men and women; for example, the negative effect of children for women but not for men; the negative effect of being Black for men but not for women. It is also worth noting that variables that relate to the area of residence are also included in the model – the wage rate and level of unemployment – as these may be expected to influence the likelihood of being in employment.

# 7.3 Logistic regression

## 7.3.1 Logit function

In the preceding sections we have discussed the use of regression with interval-level response variables. However, few variables available in census microdata (and in the social sciences in general) are measured in this way. Often we are interested in predicting whether someone has a limiting long-term illness, whether they are in paid work, or whether they have access to a car. These situations can be represented as a binary response variable, $y$, where:

- $y = 1$ indicates that the attribute is present;
- $y = 0$ indicates that the attribute is not present.

To model this variable, we want to look at the probability that the attribute is present, $P(y = 1)$. We cannot fit a linear model in this case, as the predicted outcome of the

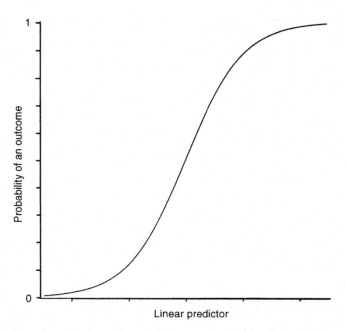

**Fig. 7.7** S-shaped transformation from a linear predictor summarizing the effects of explanatory variables to an outcome probability. (Reproduced with permission from Dale and Davies, 1994.)

linear equation

$$y_i = b_0 + b_1 x_i + e_i \tag{7.25}$$

cannot be constrained to lie between 0 and 1, which is a necessary condition if we are modelling a probability. The assumptions of OLS linear regression therefore break down. The solution is to model a function of $P(y = 1)$. The function used is known as the logit and is given by

$$P(y = 1) = \exp(z)/[1 + \exp(z)] \tag{7.26}$$

where

$$z = b_0 + b_1 x_1 + \ldots + b_n x_n \tag{7.27}$$

The logit function is plotted in Figure 7.7. Regardless of the value of explanatory variables, $x$, the outcome, $y$, is constrained to lie between 0 and 1.

The equation given above for the logit function may be rearranged to give

$$\text{Log}[P(y = 1)/P(y = 0)] = b_0 + b_1 x_1 + \cdots + b_n x_n \tag{7.28}$$

or

$$P(y = 1)/P(y = 0) = \exp(b_0 + b_1 x_1 + \cdots + b_n x_n) \tag{7.29}$$

$P(y = 1)/P(y = 0)$ (the probability of the event occurring divided by the probability of the event not occurring) gives the odds of $y = 1$, or the odds of the event occurring. As in OLS linear regression, $b_0$, $b_1$ etc. are coefficients, however the interpretation is different from that in OLS. The coefficient $b_i$ gives the change in log odds of $y = 1$ for

---

**Box 7.4 Goodness-of-fit in logistic regression**

Logistic regression does not calculate $R^2$ (although many statistical packages will compute pseudo $R^2$). For logistic regression the likelihood, $L$, is used to assess the probability of the sample of data occurring under the model. The likelihood is the product of the probabilities for each case:

$$L = \prod_{j=1}^{n} P_j(\hat{y} = y_i)$$

The higher the likelihood, the more likely it is that the sample has occurred under the given model, i.e. the better the fit of the model. The coefficients are estimated to maximize the likelihood (using maximum likelihood estimation we ascribe to the parameters the values for which the observed outcomes are most likely to have occurred if the model is correct).

As the likelihood is a product of probabilities, its maximum value is 1 (where the probability that $y = y_i$ is 1 for every case). In practice it has a small, unusable value, and hence a function of the likelihood is used instead:

$$-2 \times \log(\text{likelihood})$$

(or the −2LL). The smaller the value of the −2LL, the better the model, and a perfect model (where the likelihood = 1) will have a −2LL of 0.

---

a unit change in $x_i$, holding all other explanatory variables constant. Equivalently, $\exp(b_i)$ (the exponential of the coefficient) is the change in the odds of $y = 1$ for a unit change in $x_i$. As odds are easier to interpret than log odds, we use the second form of the equation.

Logistic regression also differs from OLS in that it does not make any assumptions about the underlying distribution of the response variable. Logistic regression is appropriate for binary response variables when the distribution of the attribute of interest lies between about 0.05 and 0.95 of the sample. From the logistic function (Fig. 7.7) one can see that this is the point at either end of the distribution where the curve flattens out. One disadvantage of logistic regression is that, unlike OLS, there is no simple measure of goodness-of-fit, although in more recent editions of SPSS a pseudo $R^2$ is computed (see Box 7.4).

Logistic regression uses individual-level data and therefore is able to deal with continuous as well as categorical response variables. This distinguishes it from logit models and loglinear models based on categorical tabular data (Demaris, 1992).

## 7.3.2 Interval-level explanatory variable

In the UK census, households are asked the number of cars or vans to which household members have access. There is a considerable body of research to show that there is a strong association between health and car access (Fox and Goldblatt, 1982, and others), that income varies directly with car access and that car access provides a reasonably efficient predictor of poverty (Gordon and Pantazis, 1997). Car access is used in England and Wales as one component of an Index of Local Conditions (Bradford *et al.*, 1993; Index Team, 1999), on which central government bases resource allocation. Whilst there is no suggestion of a causal connection between car access and health or income, and the value of having a car varies with urban or rural location and stage of life cycle, nonetheless, the number of cars in a household provides a valuable proxy for deprivation. In the following example, we consider a simple model to predict whether a household has access to a car, using census microdata from Great Britain.

A variable for whether each household has access to a car or not has been derived with the following values:

0　Has access to a car
1　Does not have access to a car

This variable is coded so that we are modelling the attribute of interest 'not having a car' – hence it is coded 1. If we were trying to explain which households were more likely to have car access then we would reverse the coding.

The data are as follows:

- *unit of analysis*: all households in GB drawn from the Household SAR for 1991 with heads aged 16 or older;
- *definition of head of household*: first named person on the census form, provided he or she was 16 or over and a usual resident;
- *variables*
  – number of cars or vans: recoded to 0/1
  – age of head of household
  – social class of head of household
  – head of household in work/not in work (work is defined as full- or part-time employment or self-employment in the week preceding the census).

As in the example for linear regression above, the first stage of fitting a model is to explore the data and look for associations between possible explanatory and response variables. In this example, however, we will start with a very simple model with one interval-level variable: age of household head (HHAGE). By including age as an interval level variable, we make the assumption that its relationship to car access is linear. In fact, as we will see later, this is not the case. Nonetheless, it is convenient to begin with this assumption.

The model equation is

$$\log \text{odds}(\text{no car/car}) = -2.57 + 0.04(\text{HHAGE}) \tag{7.30}$$

or

$$\text{odds}(\text{no/car}) = \exp[-2.57 + 0.04(\text{HHAGE})] \tag{7.31}$$

The exponential of 0.04 is 1.04, and hence the odds of not having access to a car increase by 1.04 for every unit increment in HHAGE – the older the household head, the less likely it is that the household has access to a car.

The relationship accords with our expectation that elderly heads of household will be less likely to have a car than those who are younger. In part this will be an effect of a reduced income, although there may also be a gender effect – women tend to outlive their husbands and therefore older heads of household are more likely to be women, who, in turn, are less likely than men to have a car.

To examine how well this model fits the data, we need to look at the $-2LL$. We compare the $-2LL$ of the fitted model with that of the null model, which has a constant only:

| | |
|---|---|
| $-2LL$ null model | 274 435 |
| $-2LL$ model with HHAGE | 256 309 |
| Difference | 18 126 |
| Degrees of freedom | 1 |

We can see that the $-2LL$ is lower for the model with HHAGE fitted, and hence it is a better fit. The $-2LL$ has a chi-squared distribution and, by taking the difference between the two values, we can use chi-squared tables to examine whether the improvement in fit is significant. More formally, we want to test the null hypothesis that the true population value of all the coefficients included in the model is zero. The degrees of freedom for each $-2LL$ are given by the number of cases minus the number of regression coefficients. In both models the number of cases is the same, and hence the degrees of freedom for the difference in chi-squared equals the difference in the number of coefficients included in each model. As the null model has one coefficient (the constant), and the model with age included has two (constant plus age), the difference in degrees of freedom is one.

The value for the difference, 18 126 with one degree of freedom, is significant and hence we retain age of household head in the model. However, we may note that, although significant, the reduction in the $-2LL$ is not dramatic.

## 7.3.3 Categorical explanatory variables

We expect that a household will be much less likely to have access to a car if the household head is not in paid work. Not being in a job may be due to unemployment, retirement, or, particularly in the case of female heads of household, looking after a family. The binary variable, HHWORK, was derived with the following categories:

| | |
|---|---|
| Household head in work | 0 |
| Household head not in work | 1 |

To include the binary variable, household head in work, in the model we estimate the equation:

$$\text{odds(no car/car)} = \exp(b_0 + b_i \text{HHAGE} + b_2 \text{HHWORK}) \tag{7.32}$$

When the variable HHWORK equals 0 (i.e. the household head is in work), then the model equation is given by:

$$\text{odds(no car/car)} = \exp(b_0 + b_i \text{HHAGE}) \tag{7.33}$$

When the variable HHWORK equals 1 (i.e. the household head is not working), then the model equation is

$$\text{odds(no car/car)} = \exp(b_0 + b_i\text{HHAGE} + b_2\text{HHWORK}) \qquad (7.34)$$

which is estimated as

$$\text{odds(no car/car)} = \exp(-2.07 + 0.01 \times \text{HHAGE} + 1.96 \times \text{HHWORK}) \qquad (7.35)$$

If we take the exponential and control for age, the odds of a household not having access to a car are 7.1 (exp 1.96) times higher when the household head does not have a job, compared with when the household head is in work.

Having introduced a binary response variable, we now go back to consider whether HHAGE might be better specified. Whilst we saw that as the age of the head of household increased, lack of access to a car also increased, our knowledge of households suggests that households with very young heads – under 25 – may also be less likely to have a car than those with older heads. We can therefore categorize households into three groups based on age of household head: the young – under 25s; the middle years – 25–54; older households – 55 and over. A simple exploratory analysis shows that this categorization is appropriate: 45 per cent of young households have no car, compared with 20 per cent of prime years and 50 per cent of older households.

Variables with more than two categories have to be entered into the model as dummy variables, as in linear regression. Most analysis packages do this conversion for you, but if, as in this case, you wish to group an interval-level variable, it is necessary to construct binary dummy variables by recoding. In this example our categorization is:

0   Young households (16–24)
1   Middle years (25–54)
2   Older households (55 and over)

Two dummy variables are computed with young households coded 0 in both, which thus forms the reference category:

| Age of HOH | Age category | Middle HHAGE1 | Older HHAGE2 |
|------------|--------------|---------------|--------------|
| 16–24      | 0            | 0             | 0            |
| 24–54      | 1            | 1             | 0            |
| 55 +       | 2            | 0             | 1            |

The equation is estimated as follows:

$$\text{odds(no car/car)} = \exp(-0.19 - 1.20(\text{HHAGE1}) + 0.13(\text{HHAGE2})) \qquad (7.36)$$

We can see that household heads aged 25–54 are less likely not to have a car (i.e. more likely to have a car) than the youngest age group; but that older households – 55 and over – are more likely not to have a car than the youngest age group.

To examine how well this model fits the data, we again compare the −2LL of the fitted model with that of the null model, which has a constant only:

| −2LL null model | 274 435 |
|---|---|
| −2LL model with HHAGE1 and HHAGE2 | 254 993 |
| Difference | 19 442 |
| Degrees of freedom | 2 |

We can note that the difference in the −2LL is rather larger than the model which used age as a continuous variable (difference 18 126) and therefore by categorizing age we have improved the model fit – despite reducing an interval-level variable to only three categories (although we are now fitting an extra variable to the model). In modelling this variable we might consider changing the reference category to the middle age group, thus making the reference category the one least likely not to have access to a car – the other categories would then become positive by comparison.

We have suggested that the age of the head of household and whether or not he or she is in paid work are important predictors of lack of access to a car. We may also expect that the class position of the household head will be important as this is a good indicator of spending power. The 1991 SARs coded information for current and past occupation to the Registrar General's class scheme. However, only people who had held a job in work in the last 10 years were asked for this information. Therefore the social class variable has a category which represents those with no class allocated – either because they have never had a job or because they left their last job more than 10 years ago. The groupings used here are:

| Class | | | |
|---|---|---|---|
| Upper | 0.000 | 0.000 | 0.000 |
| Middle | 1.000 | 0.000 | 0.000 |
| Lower | 0.000 | 1.000 | 0.000 |
| No class | 0.000 | 0.000 | 1.000 |

The model fitted is

$$\text{odds(no car/car)} = \exp(b_0 + b_1\text{HHAGE1} + b_2\text{HHAGE2} + b_3\text{HHWORK}$$
$$+ b_4\text{HHCLASS1} + b_5\text{HHCLASS2} + b_6\text{HHCLASS3}) \qquad (7.37)$$

Table 7.8 gives the output from fitting this model. We can see that, once the other variables are included, there is very little difference in the parameter estimates for households with heads of 55 and over and in the range 25–54. It is the reference group, age 16–24, who are least likely to have a car. The likelihood of not having a car increases as one moves down the social classes and is particularly high for households with a head who did not have a job in the previous 10 years. In some cases this will be because of age and explains the change in the sign for household heads aged 55 and over by comparison with the model where this variable was entered on its own. Not having a current job remains an important explanatory factor in lack of car access.

It is also necessary to look at the standard error terms for each parameter estimate to test the null hypothesis that the population coefficient for each term is zero. From the standard errors, we can calculate the 95 per cent confidence intervals, e.g. for the category 'middle class' the 95 per cent CI = $0.8663 \pm 1.96 \times 0.0176 = 0.90 - 0.83$. The standard output, shown in Table 7.8, also reports the significance level of each parameter estimate as well as the degrees of freedom. Analysis of census microdata such as this, where the sample size is very large, will tend to generate significant results.

**Table 7.8** Output from fitting full logistic regression model

| −2LL Null model | 274 435 |
| −2LL Model | 209 859 |
| Difference | 64 576 |
| Degrees of freedom | 6 |

| Variable household head | | B | SE | df | Sig | R | exp(B) |
|---|---|---|---|---|---|---|---|
| Age 16–24 ref | | | | | | | |
| Age 25–54 | (1) | −0.8848 | 0.0241 | 1 | 0.0000 | −0.0701 | 0.4128 |
| Age 55+ | (2) | −0.8793 | 0.0249 | 1 | 0.0000 | −0.0674 | 0.4151 |
| Upper class ref | | | | | | | |
| Middle class | (1) | 0.8663 | 0.0176 | 1 | 0.0000 | 0.0937 | 2.3781 |
| Lower class | (2) | 1.6343 | 0.0187 | 1 | 0.0000 | 0.1672 | 5.1260 |
| No class | (3) | 2.1936 | 0.0194 | 1 | 0.0000 | 0.2153 | 8.9676 |
| In work ref | | | | | | | |
| Not in work | (1) | 1.2887 | 0.0147 | 1 | 0.0000 | 0.1675 | 3.6281 |
| Constant | | −1.7915 | 0.0268 | 1 | 0.0000 | | |

The decision to include variables should therefore be based on whether we have strong theoretical reasons for their inclusion, as well as whether they are significant.

## 7.3.4 Interactions

As in linear regression we may want to fit interaction terms in our model. For example, we may expect that the effect of the head of household's social class will be different for the different age groups. The model to estimate this is:

$$
\begin{aligned}
\text{odds(no car/car)} = \exp(&b_0 + b_1\text{HHWORK} + b_2\text{HHAGE1} + b_3\text{HHAGE2} \\
&+ b_4\text{HHCLASS1} + b_5\text{HHCLASS2} + b_6\text{HHCLASS3} \\
&+ b_7\text{HHAGE1} \times \text{HHCLASS1} + b_8\text{HHAGE1} \\
&\times \text{HHCLASS2} + b_9\text{HHAGE1} \times \text{HHCLASS3} \\
&+ b_{10}\text{HHAGE2} \times \text{HHCLASS1} + b_{11}\text{HHAGE2} \\
&\times \text{HHCLASS2} + b_{12}\text{HHAGE2} \times \text{HHCLASS3})
\end{aligned}
\tag{7.38}
$$

Table 7.9 shows the output from this analysis. There are six interaction terms which correspond with the last six terms in the equation above. We can see that all these interaction terms are significant.

The −2LL for this model is 209 474, which is lower than that for the model without the interaction terms (209 859) – a difference of 385 for an additional six degrees of freedom (there are six more variables in the model with the six interactions). We can therefore conclude that the inclusion of the interaction terms does significantly improve the fit of the model. The standard errors for each term are also very small.

To interpret what the coefficients mean, we have to consider the two variables age and class of head of household. For example, the estimated equation for the reference group, age 16–24, upper class, is

$$
\text{odds(no car/car)} = \exp[-1.28 + 1.34(\text{HHWORK})]
\tag{7.39}
$$

**Table 7.9** Logistic regression model with interaction terms

| Variable | B | SE | df | Sig | R | exp(B) |
|---|---|---|---|---|---|---|
| HHAGE | | | 2 | 0.0000 | 0.0558 | |
| HHAGE(1) | −1.3062 | 0.0551 | 1 | 0.0000 | −0.0452 | 0.2708 |
| HHAGE(2) | −1.7402 | 0.0595 | 1 | 0.0000 | −0.0558 | 0.1755 |
| HHWORK(1) | 1.3361 | 0.0151 | 1 | 0.0000 | 0.1684 | 3.8041 |
| HHCLASS | | | 3 | 0.0000 | 0.0330 | |
| HHCLASS(1) | 0.3822 | 0.0615 | 1 | 0.0000 | 0.0115 | 1.4656 |
| HHCLASS(2) | 0.8902 | 0.0664 | 1 | 0.0000 | 0.0255 | 2.4355 |
| HHCLASS(3) | 1.2085 | 0.0841 | 1 | 0.0000 | 0.0273 | 3.3483 |
| HHAGE*CLASS | | | 6 | 0.0000 | 0.0368 | |
| HHAGE1*HHCLASS1 | 0.3702 | 0.0655 | 1 | 0.0000 | 0.0105 | 1.4480 |
| HHAGE1*HHCLASS2 | 0.6366 | 0.0705 | 1 | 0.0000 | 0.0170 | 1.8901 |
| HHAGE1*HHCLASS3 | 0.7181 | 0.0885 | 1 | 0.0000 | 0.0153 | 2.0505 |
| HHAGE2*HHCLASS1 | 0.8016 | 0.0696 | 1 | 0.0000 | 0.0218 | 2.2290 |
| HHAGE2*HHCLASS2 | 1.1136 | 0.0746 | 1 | 0.0000 | 0.0284 | 3.0453 |
| HHAGE3*HHCLASS3 | 1.3161 | 0.0887 | 1 | 0.0000 | 0.0282 | 3.7289 |
| Constant | −1.2773 | 0.0521 | 1 | 0.0000 | | |

For age 16–24 households in the middle class (HHCLASS1) the estimated equation is

$$\text{odds(no car/car)} = \exp[-1.28 + 1.34(\text{HHWORK}) + 0.38(\text{HHCLASS1})] \quad (7.40)$$

while for households with a head aged 25–54 in the middle class, the equation is

$$\text{odds(no car/car)} = \exp[-1.28 + 1.34(\text{HHWORK}) - 1.31(\text{HHAGE1})$$
$$+ 0.38(\text{HHCLASS1}) + 0.37(\text{HHAGE1} \times \text{HHCLASS1})] \quad (7.41)$$

We can see from the above equation that to estimate the change in odds for 25–54 year, middle class households compared with the reference category 16–24, upper class, we have to add together each component of the interaction: the term for 25–54 year households + middle class households + the interaction term. This gives $\exp(-1.31 + 0.38 + 0.37) = 0.57$.

Hence, holding householder in work constant, households with a head aged 25–54 and in the middle class are just over half as likely not to have access to a car as heads aged 16–24 in the upper class. As the interaction term is positive, we can also say that the impact of being both 25–54 and middle class is slightly more than would be expected taking each term separately – there is a small additional effect of being in both groups.

When using log odds, we add terms together in order to assess the cumulative effects of a set of characteristics, and then take the exponential to obtain the odds. Alternatively, we can take the exponential of each coefficient $\exp(\beta)$ and then estimate the overall change in odds by multiplying together the relevant terms. Using the same example as above, we can see that we arrive at the same answer:

$$0.27 \times 1.46 \times 1.45 = 0.57 \quad (7.42)$$

## 7.3.5 Conversion to probability

Often the aim of a logistic regression model applied to census microdata is to estimate or predict the probability of any particular person having a given attribute. For example, you may wish to estimate the propensity to unemployment or long-term illness. To obtain the predicted probability for any individual, the following form of the logit function is used:

$$P(y = 1) = \exp(z)/[1 + \exp(z)] \tag{7.43}$$

where

$$z = b_0 + b_1 x_1 + \cdots + b_n x \tag{7.44}$$

Using the estimated regression coefficients, $z$ is calculated from the logistic regression equation and applied to the logit function to give an estimate of the probability of the event occurring or of possessing a given attribute.

For example, from the model with head of household's age group, work position and class fitted (Table 7.8), the probability of a household with a head aged 55+, in the working class and not in a paid job having access to a car is given by

$$z = [-1.79 - 0.88(\text{HHAGE2}) + 1.29(\text{HHWORK}) + 1.63(\text{HHCLASS2})]$$
$$z = 0.25 \tag{7.45}$$

(Note that all three dummies in the above equation are equal to 1.)

$$P(\text{no access to a car}) = \exp(0.25)/[1 + \exp(0.25)]$$
$$= 0.56 \tag{7.46}$$

If we recalculate the probability, substituting upper class (the reference category) for working class, our equation becomes:

$$z = [-1.79 - 0.88(\text{HHAGE2}) + 1.29(\text{HHWORK})]$$
$$z = -1.38 \tag{7.47}$$

and the probability falls to

$$P = 0.2$$

## 7.3.6 Alternative modelling techniques

Logistic regression is a widely used technique in social science research. However, the researcher should be aware that there are a variety of techniques available, the choice of which will depend both on the data structure and on the research questions asked.

For example, it may not be possible to define a dependent variable as binary (such as access to car/no access to car). In a model of women's employment, it may be more appropriate to distinguish between full-time employment, part-time employment and no paid employment. In this example, it is not possible to fit a straightforward logistic regression model as the response variable is not binary but multinomial. A multinomial response variable (with three or more categories) can be modelled using multinomial logistic regression, available in SPSS (version 9 and above), STATA and CATMOD in SAS. In a multinomial logistic regression, the response variable is assumed to be categorical and one obtains two sets of comparisons from which the third can be derived. Thus, in the example above, one would model the odds of being in full-time work versus part-time work and in full-time work versus not being in work. From this we can obtain the third combination – the odds of part-time working versus not being in work. A detailed discussion of multinomial logistic regression is given by Demaris (1992).

Sometimes we have a response variable with three or more categories but with a clear ordering, e.g. with educational qualifications or social class. Where this occurs, the ordered nature of the response variable is modelled as an ordered probit or ordered logit. This analysis is very straightforward to run in STATA and SAS.

The probit transformation uses a normal distribution as opposed to a logistic distribution used in logistic regression. Probit models give very similar results to logit models, but the results of the latter are usually assumed to be easier to understand. However, probit models are widely used where selection bias needs to be taken into account.

Loglinear models are a useful way of analyzing multi-way tables, e.g. where one wants to examine the relationship between access to cars, housing tenure, social class and gender. They do not require a dependent variable to be defined and therefore make no assumptions about causality. They are widely used in the analysis of inter- and intra-generational social mobility (Raftery, 1985; Clifford and Heath, 1993; Breen and Hayes, 1996; Krymkowski et al., 1996) where very complex models have been developed. In the analysis of census microdata, loglinear models may be particularly valuable when one wants to establish whether a relationship between, say, education and social class has remained constant over time, or across countries. A detailed discussion of loglinear and logistic models for tabular data is given in Gilbert (1993).

## 7.3.7 Modelling the clustered nature of observations

Multivariate regression analysis and logistic regression are both widely used in the analysis of surveys and samples of census microdata. They both assume that a simple random sample has been taken and ignore any clustering that may be present in the data. One of the strengths of samples drawn from a census of population is that there is no requirement to use clustering in order to reduce the costs of data collection. Even though the sampling design may not contain any clustering, clustering may still occur naturally due to the way in which the data are structured. Where this occurs, there are good reasons to use methods of analysis which recognize this clustering (Skinner et al., 1989). Multilevel modelling has been developed for this purpose and is discussed in the following section.

## 7.4 Multilevel modelling

Data may be naturally structured in a number of ways. For example, within the field of educational research it has long been recognized that the achievement of school pupils is related to their classroom teacher, to the characteristics of their school, and to the district (catchment area) from which they are drawn. Thus there is a correlation between pupils who share the same teachers; pupils who go to the same school; and pupils who live in the same neighbourhood. If this structure is ignored, then explanatory models of educational attainment may be quite misleading (Goldstein, 1995). Using a multilevel approach, it is possible to simultaneously model compositional effects (individual-level characteristics) and contextual effects (ecological or area-level influences).

Analysis of microdata has shown that it is important to allow for geographical clustering in the population, which is related to a number of socioeconomic characteristics such as unemployment (Gould and Fieldhouse, 1997; Fieldhouse and Gould, 1998; Tranmer and Steel, 1998). Methods for doing this are explored in the following section.

### 7.4.1 The structure of census microdata

Because of its large sample size design, census microdata can provide a reasonable degree of geographical precision about where respondents live. In this respect, it differs from survey data which tend to lack geographical detail, normally because the sample size is not sufficiently large to allow for geographical disaggregation. Census microdata, although constrained by the need to protect confidentiality, commonly allow areas with populations of around 100 000 to be identified (see Ch. 2). This means it is possible to take account of where a person lives when modelling his or her characteristics. For example, when modelling the propensity to illness, it may be important to allow for geographical differences, and the impact of 'contextual' or environmental variables (see Section 8.5).

Census microdata can be viewed as being hierarchically structured, with each individual living within a particular geographical area. Typically, individuals within these higher level units (areas) are more homogenous with respect to the item of interest than the population as a whole. In most traditional statistical modelling procedures, this correlation is a problem, and error terms may be underestimated. One way of avoiding this problem is by using multilevel modelling techniques. Multilevel logit models provide an efficient way of allowing individuals and the areas within which they live to be modelled simultaneously (Section 8.5; Gould and Fieldhouse 1997; Fieldhouse and Gould 1998).

Multilevel models therefore exploit the hierarchical structure of census microdata, measuring the degree of similarity within and between areas rather than treating correlation of individuals as a nuisance factor (Goldstein, 1995). Because multilevel models can simultaneously model both the micro (individual) and the macro (area) level, it is possible to assess whether area or place differences are important having taken into account the characteristics of individuals (compositional variables)

(Jones and Duncan, 1995). Thus multilevel models allow the effect of 'place' to be analyzed net of the 'confounding' effect of individual and household characteristics such as age, gender, ethnicity, marital status, housing tenure and social class. It is also possible to test whether factors affecting the response variable vary between places, and whether there is differential variability between groups. For example, is there a different geography of unemployment for the different ethnic groups, and if so, is one group more spatially variable than another (Fieldhouse and Gould, 1998)? Area-level variables can be added to the model to explain the variation observed between different areas, after individual variation has been accounted for.

The British SARs allow an extension of the multilevel geographical structure through the addition of area classifications. The 1991 Individual SAR for GB contains a descriptor of the enumeration district (ED) in which each person was resident, attached to each record. Although this does not identify the name of the ED in which an individual lives, it provides valuable contextual information that describes the person's neighbourhood. If social context is assumed to be important in affecting people's life chances, then this contextual descriptor can be built into a geographical model in a number of ways, the most obvious being a 'cross-classified' model where individuals are nested within identified SAR areas and also within types of neighbourhood, identified by the area classification (Fieldhouse and Tranmer, 1999). The models described below do not use an area classification and are straightforward hierarchical models.

## 7.4.2 Visualizing a multilevel model

Where continuous data are being modelled, diagrammatic representation of a multilevel model is relatively simple. Because of the lack of continuous-level variables in census microdata, it is helpful to illustrate a multilevel linear model using other data sources. Figure 7.8(a) from Plewis (1994) shows the relationship between pupils' attainment and a measure of family income. This assumes the same relationship for all pupils regardless of the school they attended. However, in reality, pupils from some schools are more likely to perform well than pupils from other schools, and this may be partly attributable to a school-level effect (e.g. quality of teaching). A multilevel analysis allows us to separate the pupil- and school-level effects. Following this, Figure 7.8(b) shows three separate but parallel regression lines, representing one for each school. Whilst this allows for a different average level of attainment within each school, it still assumes the same relationship between attainment and family income in each school. In other words, the intercept may be different but the slopes do not vary. In Figure 7.8(c) are allowed both the intercept and the slope to be different. In other words, the relationship between income and exam performance is different for each school. The best-fit regression lines for the three schools in this example are therefore not parallel. By modelling the natural structure of the data, it is possible to avoid misleading inferences and include a much clearer analysis of the different effects which influence a child's educational attainment (Raudenbush, 1993; Goldstein, 1995).

Census microdata consist largely of nominal variables for which linear models are inappropriate. The following is based on an analysis of geographical variation in

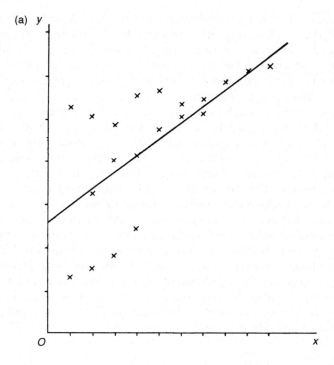

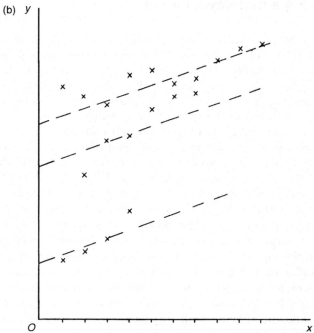

**Fig. 7.8** (a) OLS regression line; (b) Variance component model: random intercept. (Reproduced with permission from Plewis, 1994.)

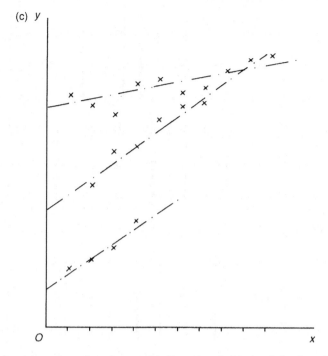

(c) *y*

*O*                                                                          *x*

**Fig. 7.8** (c) Random intercept, random slope model. (Reproduced with permission from Plewis, 1994.)

long-term limiting illness using data from the 1991 2 per cent SAR for Britain. The modelling strategy discussed is a binary logit multilevel model and can thus be seen as an extension of the logistic regression of the previous section. The results are discussed in Section 8.5. Using a binary response variable, multilevel models can be described in terms of a series of hypothetical graphs which show 'place' differences in levels of limiting long-term illness (Jones and Duncan, 1996). In each part of Figure 7.9 the vertical axis represents the probability of an individual experiencing a long-term limiting illness, while the 'columns' on the horizontal axis represent sex.

Figure 7.9(a) represents a 'fixed-effects' model; that is, the results that would be obtained if the usual single-level model were fitted to the data on limiting long-term illness (LLI). The model assumes the same national rates of LLI for men and women, thereby denying any geographical variation. In other words, LLI rates are 'fixed' at their national rate and not allowed to vary between places. This is equivalent to a single-level logistic regression model with one explanatory variable (sex). It also corresponds to Figure 7.8(a), in that it is a single-level model with one overall relationship between sex and illness for all SAR areas.

Figure 7.9(b) represents a random intercepts or 'variance components' model, which corresponds to Figure 7.8(b). It is a simplified schematic version with just six SAR areas. This graph retains the national relationship (the fixed effect) for LLI shown by the thicker line, but now the average level of long-term illness is allowed to vary from place to place. The two columns representing males and females contain

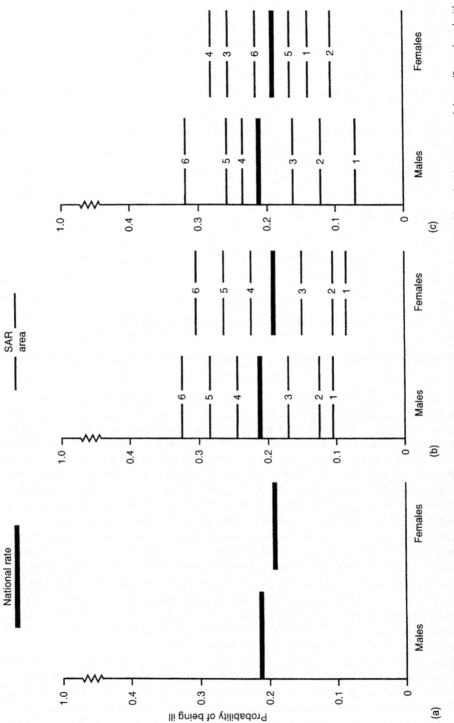

**Fig. 7.9** Multilevel models with varying relationships: (a) fixed effects; (b) random intercepts or 'variance components'; (c) random intercepts and slopes. (Reproduced with permission from Gould and Jones, 1996.)

a series of separate (thinner) lines that indicate the different probabilities of LLI for different SAR areas. The differential between SAR areas, however, is identical for men and women. For example, notice that area 6 has the highest rate for men and women. To put it another way, there is now a geography, but it is a simple one which could be represented by a single map; places with elevated rates for men also have higher rates for women.

Finally, we wish to consider the situation described in Figure 7.8(c) which allowed for a different relationship between the explanatory and response variables for each level 2 unit (SAR area in Fig. 7.9). The model contains two sets of random terms, one for each sex (Fig. 7.9(c)). This figure indicates that the place-specific probabilities are different for men and women; rates for illness are both higher in general and more variable for men then for women. There is not one single map of gender differences, but complex between-place heterogeneity. SAR area 6 has the highest probability of LLI for men but only the third highest for women. The aim of multilevel modelling is to distinguish between these different models, to assess whether place differences are important, and whether they are relatively simple or complex, and to do this taking account of age, ethnicity, class and other relevant individual-level characteristics.

## 7.4.3 Exploiting the hierarchical multilevel nature of census microdata

Table 7.10(a) illustrates some individual records taken from the 2 per cent Individual SAR for GB. The table shows the response to the long-term illness question, three individual level 1 variables (age, ethnicity and gender) and the code number of the SAR area in which the individual resides. These individual-level records can be easily aggregated into the cross-tabular format shown in Table 7.10(b) (Jones, 1994).

**Table 7.10** Limiting long-term illness

(a) **Age, ethnicity and gender as individual records**

| Respondent | Long-term illness | Age | Ethnicity | Gender | SAR area |
|---|---|---|---|---|---|
| 1 | Yes | 35 | White | Male | 1 |
| 2 | No | 52 | White | Male | 1 |
| 3 | Yes | 26 | Black African | Female | 1 |
| 419 550 | No | 46 | White | Female | 278 |

(b) **Proportions by age, gender and ethnicity for SAR area 1**

| | Males | | Females | |
|---|---|---|---|---|
| Age | White | Black | White | Black |
| 30–39 | 2/182 | 7/53 | 7/183 | 5/49 |
| 40–49 | 13/154 | 4/28 | 19/177 | 1/38 |
| 50–60 | 18/144 | 111/36 | 34/169 | 7/83 |

[*Source:* Individual SAR. Crown copyright]

Table 7.10(b) represents an aggregation of all the individual records for SAR area 1, (City of London and City of Westminster), with each cell in the table giving the ratio of the numerator (number of ill people) to the denominator (total number of people). It is these proportions that are modelled as the response. Each SAR area in the analysis will have a similar table, with the data organized so that 'cells' (level 1) are nested within places (level 2). There need be no loss of information in going from Table 7.10(a) to Table 7.10(b), although in this example age and ethnicity have been re-categorized in the latter to reduce excessive sampling fluctuations. Recasting the data in this form may reduce the number of level 1 units that have to be modelled and thereby reduce the computation time for the analysis (Wong and Mason, 1984).

More individual-level variables (level 1) can be added, creating a larger, multi-way table with many more dimensions than Table 7.10(b). For example Gould and Jones (1996) undertook an analysis of long-term illness with the following variables and categories: sex (2), age (3), social class (4), housing tenure (3), car ownership (3), ethnicity (2) and 278 SAR areas (see Section 8.5). The maximum possible number of cells in the cross-tabulation of variables and categories was therefore 120 096. Many of these cells (about 23 per cent) were 'empty', as there were no individuals with particular combinations of characteristics in particular SAR areas. In this example, creating the data structure as cells within places resulted in a nearly eightfold reduction in the number of level 1 units for computation and model estimation. Recoding data, as in this example, represents a balance between retaining detailed information on individuals and ensuring that disaggregation does not lead to excessive sampling fluctuations in cell counts. There is always a need to generalize and simplify large datasets, and multilevel models provide one method that can simultaneously give useful sociological and geographical insights.

## 7.4.4 Specification of multilevel models

The model discussed here uses a binary response variable, defined as the proportion of people in a cell who experience limiting long-term illness with respect to the total number of people in the cell. The explanatory variables are individual- or house-hold-level characteristics (ethnic group, gender, age, social class, access to a car and tenure), set up as a series of dummy variables. In this respect, the data could equally well be used in a logistic regression analysis, as discussed in the previous section. The only difference is that, in this example, we also have a variable that indicates the area of residence. The fact that the data are organized with the response variable expressed as a proportion in each cell of the cross-tabulated explanatory variables is simply a device to reduce the amount of computation required.

In the example used here, the base or reference category represents White males aged 30–40, living in owner-occupation, social class III with one car, and is signified by the 'constant', while the other 'individual' characteristics are specified as a set of contrasts, with a 1 given if a cell belongs to a particular characteristic, and 0 otherwise. In the following equations, we adopt the notation that is commonly used in multilevel texts (e.g. Goldstein, 1995) as it provides a framework with which to represent the greater complexity of the models.

## Model 1: fixed effects

We begin by considering the fixed-effects micro-model (i.e. the model for individuals) illustrated in Figure 7.9(a). This model can be expressed as[1]

$$y_{ij} = p_{ij} + e_{ij} \qquad (7.48)$$

where $y_{ij}$ is the LLI status of individual $i$ in area $j$, $e_{ij}$ is the error term for individual $i$ in area $j$ (the level 1 residual) and $p_{ij}$ is the predicted probability of individual $i$ in area $j$ having a LLI, where the logit of $p_{ij}$ can be expressed as a function of the individual-level explanatory variables as follows:

$$\text{logit}(p_{ij}) = \beta_0 + \sum_{m=1}^{r} \beta_m x_{mij} \qquad (7.49)$$

where $x_1$ to $x_r$ are a set of explanatory variables (e.g. sex); $\beta_0$ is the constant term (in this case the proportion of 30–39 year-old White males who are owner occupiers in social class III owning one car and reporting illness); and $\beta_1$ to $\beta_m$ are the coefficients for each explanatory variable (in this case the contrasts in the proportion who are ill between the other demographic and socioeconomic groups).

The model simply states that the proportion of people experiencing LLI is a function of the individual characteristics (explanatory variables), and that this does not vary from place to place. The model contains only parameters specified at the individual level, and therefore presumes that there is no geographical variation in perceived levels of LLI. It is therefore directly comparable to the logit models discussed in Section 7.3.1.

## Model 2: random intercepts

Geographical variation is introduced by specifying a random intercepts model, as illustrated in Figure 7.9(b). In other words, we assume there is a fixed relationship between explanatory variables and the response variable across all areas, but allow the average level to be different in each SAR area. Now $p_{ij}$, the predicted probability of individual $i$ in area $j$ having a LLI, includes a level 2 error term $\mu_{0j}$, which represents the difference between the average level of illness for area $j$ and the overall mean (i.e. $\mu_{0j}$ is the level 2 residual). The logit of $p_{ij}$ can now be expressed as

$$\text{logit}(p_{ij}) = \beta_0 + \sum_{m=1}^{r} \beta_m x_{mij} + \mu_{0j} \qquad (7.50)$$

The level 2 random term, $\mu_{0j}$, represents the place-specific differences in the rate of LLI. The distribution of these differences is summarized in a multilevel model by a level 2 variance term: $\sigma_{\mu_0}^2$. If this term is not significantly different from zero, there are no overall place differences, and the variations in illness can be adequately

---

[1] Because we have fitted the response as a proportion instead of as a dichotomous variable for each individual, the assumption of binomial variance no longer holds, since the proportions are based on the sum of separate binomial variables with differing probabilities. To take this into account, an extra binomial parameter can be fitted at level 1 (Goldstein, 1995, p. 98). The default settings on MlwiN constrain this term to one (i.e. assume binomial variance) and therefore in this case (model 1) we have excluded it from the equation (see also Diamond et al., 1999). In some cases (e.g. where the model is not properly specified) the assumption of binomial variation should be relaxed.

summarized by individual-level characteristics. In other words, if $\sigma^2_{\mu_0}$ is zero, a multi-level model is not required, as each area has the same level of long-term illness.

### Model 3: random intercepts and slopes

Model 2 can be extended to allow LLI not only to vary between SAR areas but also to vary with individual characteristics – in this case sex. This in shown in Figure 7.9(c) and is expressed as

$$\text{logit}(p_{ij}) = \beta_0 + \sum_{m=1}^{r} \beta_{mj} x_{mij} + \mu_{0j} \tag{7.51}$$

In model 3 we have allowed the relationship between LLI and the explanatory variables to be different in each area, by including area-specific parameters, $\beta_{mj}$. These may be defined in general, for $m = 1, \ldots, r$ explanatory variables, as

$$\beta_{mj} = \beta_m + \mu_{mj} \tag{7.52}$$

$\beta_{mj}$ is the overall relationship between explanatory variable $m$ and the response variable, LLI, and $\mu_{mj}$ represents the deviation in this coefficient for area $j$ (when there is only one explanatory variable, as in Fig. 7.9(c), we can write $\beta_{1j} = \beta_1 + \mu_{1j}$). In other words, the addition of this random parameter at level 2 allows there to be a different relationship between sex and LLI in each area. If variance of $\mu_{mj}$ ($\sigma^2_{\mu_m}$) is equal to zero, the relationship is the same across all areas, although the level may vary.

In the example portrayed in Figure 7.9(c), there is one explanatory variable, sex, a dummy variable ($x_1$) which takes the value of 1 for women and 0 for men. There are therefore two sets of random terms that correspond to sex: $\mu_{0j}$ gives the place-specific difference for the base category (men), and $\mu_{1j}$ represents the place-specific differentials for women. The two sets of random terms at level 2 are summarized by two variances and a covariance ($\sigma^2_{\mu_0}$, $\sigma^2_{\mu_1}$ and $\sigma_{\mu_0\mu_1}$). Hence the between-place variation for men will be given by $\sigma^2_{\mu_0}$ and for women by $\sigma^2_{\mu_0} + 2\sigma_{\mu_0\mu_1} + \sigma^2_{\mu_1}$.

In order to convert the predicted values ($p$) to probabilities, it is necessary to apply the exponential transformation, in exactly the same way as for a single-level logistic regression model as described in Section 7.3.5.

The models discussed here can be further extended to allow for under- or over-dispersion in the response variable (extra binomial variation). For a detailed discussion of these issues, see Goldstein (1991), Wrigley (1985), Jones (1994), and Collett (1991). The models may also be extended to allow for differential variability in the response variable for people with different characteristics (Bullen *et al.*, 1997).

The modelling techniques described here are applied in Section 8.5, which provides an empirical example of a multilevel model of long-term illness using the 1991 UK Individual SAR.

## 7.4.5 Further applications

The preceding section has described the application of two-level models to census microdata. However, some census microdata allow for more complex multilevel models to be fitted. These include the use of area-level classifications attached to microdata, and the analysis of the hierarchical element of household data.

## Area classifications

The recent inclusion in the SARs of an area-level classification which provides a descriptor of the type of area in which an individual lives (Section 2.5.5) opens up exciting opportunities for model refinement and the analysis of geographical variations. These are potentially very useful as they provide an opportunity to assess the relative importance of two non-hierarchical overlapping sets of contextual units in accounting for variations in morbidity. This can be achieved by fitting cross-classified extensions of strictly hierarchical multilevel models where individuals at level 1 are cross-classified by SAR areas (geographical contexts) and GB profiles (residential contexts), both at level 2 (Goldstein, 1995; Jones et al., 1998). For example, Fieldhouse and Tranmer (1999) investigate the variation in unemployment between individuals, ED types and SAR areas. They also examine a similar variation in deprivation. The addition of the area-level classification to the SARs opens up new possibilities for multilevel modelling, by combining these data with tabular data (small area statistics) from the main census.

## Household analysis

The 1 per cent Household SAR provides opportunities to explore household variations and assess the extent to which individual characteristics are clustered amongst individuals living in the same household (Dressler, 1994). Whilst the Household SAR has a hierarchical structure where 538 819 individuals are nested within 214 369 households, which are, in turn, nested in 10 standard regions in England, plus Wales and Scotland, the small number of regions probably precludes their inclusion as units in a three-level model. However, regions could be used as dummy variables in a two-level model with a fixed part expansion (Jones and Bullen, 1994). Moreover, now that a ward-level classification has been added to the 1 per cent Household file, these 'residential types' could be used as part of a higher-level contextual unit in a three-level model or again as a dummy variable in a fixed part expansion of a two-level model.

## Appendix 7.1: STATA commands for modelling the selection process using Heckman's procedure

Heckman's correction for sample selection can be readily applied in STATA using standard commands given in the manual when the selection equation is modelled with a binary response variable. However, a multinomial response is not available as a standard command. We are grateful to Phil Murphy, University of Swansea, for providing the equations, given below, for modelling a participation probit with a multinomial response.

## Multinomial

### *Modelling a participation probit with a multinomial outcome term in STATA*

Multinomial logit specification:

$$\Pr(I_s = 1) = \left\{ \exp(M_s\pi_s) \middle/ \left[ \sum_{k=1,2,\dots,m} \exp(M_k\pi_k) \right] \right\} \qquad \text{for } s = 1, 2, \dots, m$$

Selectivity correction terms (Lee, 1982):

$$\lambda_{1s} = \left\{ \phi[J_s(M_s\pi_s)]/\Pr(I_s = 1) \right\} \qquad s = 1, 2, \dots, m$$

where $\phi$ is the standard normal density function; and $J_s(M_s\pi_s) = \Phi^{-1}[\Pr(I_s = 1)]$ is the inverse function of the cumulative normal distribution.

The problem in STATA is finding the inverse of the cumulative normal distribution. However, this can be approximated by

$$H1(s) = \frac{U1(s) \times (2.5101 - 12.2043 \times U1(s)^2 + 11.2502 \times U1(s)^4)}{(1 - 5.8742 \times U1(s)^2 + 7.7587 \times U1(s)^4)}$$

where $H1(s) = J_s(M_s\pi_s)$; and $U1(s) = \Pr(I_s = 1) - 0.5$, which is derived from the predicted probabilities from the multinomial logit model.

The selectivity terms are then found as the ratio of the standard normal to cumulative normal evaluated at $H1(s)$.

# Exemplars of analyses of census microdata

## 8.1 Introduction

In this chapter we use examples from specific analyses conducted by experts in the area to exemplify some of the features of census microdata and some of the analytic techniques discussed earlier. Each section of this chapter covers a different topic.

Section 8.2 focuses on the unique role of the census in enumerating the entire population, not just those in private households. The authors show how selection effects may lead to misleading assumptions about the relationship between marital status and health if elderly people in institutions are omitted from analyses. (The coverage of the census is discussed in Chapters 1 and 2.)

Section 8.3, by Suzanne Model, provides an example of cross-national comparison based on Britain, the USA and Canada and thus exemplifies many of the issues raised in Chapter 3. It highlights some of the difficulties of making comparisons when the topics of interest are asked in very different ways in the three countries. It also offers some innovative suggestions for overcoming these difficulties. In addition, it demonstrates the value of census microdata in providing a large enough sample size to focus on one relatively small subgroup of the population – the Chinese.

Section 8.4, by Clare Holdsworth, also makes a cross-national comparison between Britain and the USA, which uses questions on ethnic group and race. Whilst Model, in Section 8.3, based her comparisons on a relatively well defined group – the Chinese – Holdsworth discusses some of the challenges the analyst faces when making comparisons across a wider range of subpopulations. However, this section also provides an important exemplar of the value of hierarchical data to study the extent to which couples share the same racial origin or ethnic group and to analyze the relationship between the ethnic/racial group of children and that of their parents. The methodological details are discussed in Section 5.3. Again, census microdata provide a rare opportunity to analyze the detailed family formation of small groups within the population.

The value of census microdata for area-level analysis forms the focus of the last three sections of this chapter. In Section 8.5, Myles Gould and Kelvyn Jones describe a multilevel analysis of the variation in limiting long-term illness not only by individual characteristics such as ethnic group, but also by place. This highlights one of the other unique characteristics of census microdata – geographical definition which is much finer than available on standard social surveys.

In Section 8.6 Ed Fieldhouse demonstrates how an indicator of deprivation can be derived for individuals and use this to show the extent to which 'deprived' individuals live in 'deprived' places, thus providing another example of the value of the geographical detail in census microdata.

In Section 8.7, Stephen Simpson gives an overview of ways in which census micro-data can be used to improve small-area estimation. In all cases, this requires combining microdata with other census data or with survey data. The ability to link together data sources opens up many exciting possibilities and this section ends by hinting at some of the future possibilities. The methods for these examples are not covered in earlier chapters but each is referenced for further reading.

## 8.2 The importance of institutional populations in analyses of health in later life
### Emily Grundy, Karen Glaser and Mike Murphy

The research reported here was supported by the Economic and Social Research Council as part of its Population and Household Change Programme, grant L31523018, and is based on analyses of the Samples of Anonymised Records from the 1991 census, provided by the Census Microdata Unit at the University of Manchester with support of the ESRC and JISC.

### 8.2.1 Introduction

This chapter draws on detailed analyses of relationships between health and household type (Glaser et al., 1997b; Grundy and Glaser, 1997; Murphy et al., 1997) to illustrate the value of datasets such as the SARs which allow analyses based on the whole population, rather than, as in most sample surveys, the private household population only. Numerous studies have shown that individuals in institutions have poorer health than those in private households and attention has been drawn to the bias attendant on excluding the institutionalized population from analyses of trends and variations in health statistics (Grundy, 1992a; Bone et al., 1996; Ebrahim, 1996). However, few studies have examined the extent of this bias in detail. Omitting this considerably less healthy group from analyses of health in later life is important for several reasons. Although the overall proportion of individuals in institutions is small, the percentage of the oldest old in communal establishments is quite high in many Western countries and their exclusion may bias estimates of illness rates for this age group. Secondly, there are strong associations between variables such as marital status and likelihood of residence in an institution (Verbrugge, 1979; Börsch-Supan, 1990; Burr, 1990). Never-married individuals and those not currently married are more likely to enter institutions than the married (Grundy, 1992b; Grundy and Glaser, 1997). Excluding individuals in communal establishments may therefore affect estimates of relative health status by marital status. The focus of this section is the impact of institutional populations on estimates of health in later life, and how inclusion of this group may inform the debate on the relationship between health and marital status at older ages.

## 8.2.2 Empirical associations between marital status and health among the elderly

There is conflicting evidence with respect to the association between marital status and health in later life. In contrast to the conventional pattern of better health among the married compared to the never-married, Goldman *et al.*'s (1995) study, based on a national sample of the private household population (the US Longitudinal Study of Aging), found never-married older women to have better health outcomes than their married counterparts, a result the authors attributed to their supportive social environment.

This analysis uses the 2 per cent Individual SAR for Great Britain, which includes the population in communal establishments, to examine how inclusion of the institutionalized population affects both health estimates in later life and the relationship between health and marital status.

## 8.2.3 Data and methods

There are several advantages to using the 2 per cent SAR to examine the health status of the elderly population. In comparison to surveys like the General Household Survey, which is based on a much smaller sample size (around 10 000 households each year), the SARs are much larger: the 1 per cent sample is 20 times larger and the 2 per cent sample 40 times larger. This permits detailed analyses of the elderly population, either by single year of age or the older age group (85 and older). As already explained, the SARs also allow inclusion of the institutionalized population – an often omitted but important group for studies examining the health status of older people.

In this analysis, the private household population includes usual residents, but 'visitors' – those present on census night but not usually resident – are excluded. The non-private household population is defined as those individuals enumerated in non-private establishments where 'some form of communal catering is provided' (OPCS and GRO(s), 1992). This broad definition covers a wide range of institutions, such as medical and care establishments, detention, defence and education establishments, hotels, boarding houses, hostels, common lodging houses, and other categories of non-private households (persons sleeping rough, campers, and those on civilian ships, boats and barges) (OPCS and GRO(s), 1992). Our analysis excludes visitors in communal establishments, but includes both resident staff and non-staff, i.e. those individuals who stated 'this address' as their usual address.[1] Health status is assessed using the census question on limiting long-term illness.

## 8.2.4 Results

### *The institutional population*

Although under 2 per cent of men over 45 and about 3 per cent of women in this age group are in institutions, the proportion increases sharply at older ages, so that for

---

[1] In addition, information for those households with 12 or more people was not available in the 1 per cent SAR for reasons of confidentiality.

**Table 8.1** Distribution of adults aged 85 and over in 1991 resident in institutions by type of institution[a]

| Type of institution | ($N = 4045$) |
|---|---|
| Psychiatric hospitals/homes (NHS or non-NHS) | 2.3 |
| Other hospitals/homes (NHS or non-NHS) | 5.3 |
| Local authority homes | 22.6 |
| Housing association homes | 2.2 |
| Nursing homes | 29.2 |
| Residential homes | 37.2 |
| Other | 1.2 |
| Total | 100 |

[*Source*: 2 per cent Individual SAR for GB]
[a] The number in institutions includes only non-staff residents.

those aged 85 and over, 15.6 per cent of men and nearly a third of women are in institutions (Murphy *et al.*, 1997). The oldest old make up a substantial proportion (over one-quarter) of the institutionalized population. Gender differences among the institutional populations are remarkable: only 10 per cent of men in institutions are aged 85 and over compared with 40 per cent of women (2 per cent SARs). Over three-quarters of residents aged 85 and over in non-private households are in local authority, nursing and residential homes (Table 8.1).

## Estimates of health: including the institutionalized elderly

In the older age groups, the number in non-private households account for a substantial proportion of all those reporting limiting long-standing illness, especially among women, so that the omission of this population will bias health estimates at older ages (Fig. 8.1). The importance of including the institutionalized population in estimates of the frequency of illness in the population, especially for the older age groups, can be seen in Table 8.2 which compares long-term illness rates for residents in private households and communal establishments. Long-term illness rates are much higher for the 85 and over age group – especially amongst women – when the population in communal establishments is included. The difference in limiting-long term illness rates among these older age groups between the private household population and the total population is more marked for single and widowed individuals than for those in the other marital status categories, reflecting the greater likelihood of entry into institutions of the single and widowed (Table 8.2). Surveys like the General Household Survey, based only on private households, underestimate the prevalence of illness in the elderly population.

## The relationship between health and marital status at older ages, including the institutionalized

The probability of residing in an institution is clearly associated with marital status, as well as with health, which is likely to affect the prevalence of illness among various marital status categories at older ages. In the private household population, a large proportion of the younger age groups are single, and this rises in the married/remarried and divorced/widowed groups as the population ages. In contrast to the private household population, the marital status distribution of the non-private household population shows large numbers of widowed and single individuals, especially in the older age groups (Fig. 8.2). The startling difference in the

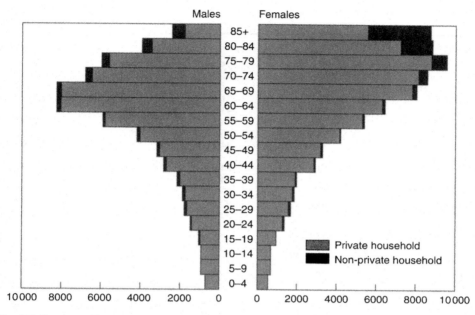

**Fig. 8.1** Numbers of long-term sick by type of residence, GB 1991. [*Source*: 1991 2 per cent Individual SAR]

proportions of single versus married people in institutions may partly reflect either errors in enumeration, if those in charge of communal establishments are more likely to list the residents as single, or the fact that married individuals in institutions may be more likely to list their home addresses as their usual residence. While cross-sectional data such as the SARs cannot show patterns of movement into institutions, longitudinal analysis for the period 1971–81 (Grundy, 1992b), showed that the great majority of older single people in institutions in 1981 had been in the private household sector 10 years earlier and that marital status was a significant factor associated with transitions to non-private households. Thus, the exclusion of the institutionalized population among the elderly will lead to the underestimation of the prevalence of a limiting long-term illness and distort the true relationship between marital status and the prevalence of such illness. For example, at ages 85 and over, among single people with limiting long-standing illness, nearly half are in institutions, compared with only about 15 per cent of the equivalent married population (Table 8.3).

Figure 8.3 shows differences in the prevalence of a limiting long-term illness between specified marital status groups and the overall percentage value for women in the same age group, in the private household sector (Fig. 8.3a) and in the total population (Fig. 8.3b). It can be seen that among women aged 40 and over in private households, until about age 75, those in their first marriage report the lowest rates of limiting long-standing illness, while for those under 55, single women report the highest prevalence (Fig. 8.3a). Beyond age 75, however, single women in private households show the *lowest* levels of reported limiting long-standing illness, while the divorced group generally has the highest values.

**Table 8.2** Adults aged 65 and over in 1991 with a limiting long-term illness (%) by age, sex and marital status

(a) **Men**

|  | 65–74 | 75–84 | 85+ | 65+ |
|---|---|---|---|---|
| Private households |  |  |  |  |
| Single | 37.8 | 41.2 | 47.8 | 39.1 |
| Married | 33.5 | 43.6 | 58.0 | 37.2 |
| Divorced | 38.6 | 46.3 | 51.2 | 40.4 |
| Widowed | 38.3 | 46.8 | 58.1 | 44.8 |
| All | 34.6 | 44.3 | 57.4 | 38.8 |
| Institutions[a] |  |  |  |  |
| Single | 84.4 | 84.6 | 97.6 | 86.3 |
| Married | 84.3 | 94.7 | 94.2 | 91.9 |
| Divorced | 86.0 | 83.7 | 87.5 | 85.2 |
| Widowed | 85.6 | 92.3 | 94.3 | 92.2 |
| All | 84.8 | 90.7 | 94.6 | 90.2 |
| Total[a] |  |  |  |  |
| Single | 42.1 | 47.9 | 63.5 | 44.9 |
| Married | 33.7 | 44.4 | 60.5 | 37.8 |
| Divorced | 40.1 | 50.1 | 56.9 | 42.6 |
| Widowed | 39.7 | 50.3 | 65.8 | 48.6 |
| All | 35.3 | 46.3 | 63.2 | 48.6 |
| N | 43 286 | 21 914 | 4028 | 69 228 |

(b) **Women**

|  | 65–74 | 75–84 | 85+ | 65+ |
|---|---|---|---|---|
| Private households |  |  |  |  |
| Single | 31.3 | 42.1 | 59.6 | 39.5 |
| Married | 29.5 | 45.8 | 61.3 | 34.5 |
| Divorced | 34.7 | 48.4 | 66.7 | 39.2 |
| Widowed | 32.4 | 47.8 | 64.5 | 44.0 |
| All | 30.9 | 46.7 | 63.5 | 39.6 |
| Institutions[a] |  |  |  |  |
| Single | 82.7 | 92.2 | 92.9 | 90.8 |
| Married | 94.6 | 93.0 | 94.4 | 93.8 |
| Divorced | 92.9 | 89.3 | 77.3 | 88.0 |
| Widowed | 92.7 | 94.5 | 95.3 | 94.8 |
| All | 89.6 | 93.9 | 94.7 | 93.8 |
| Total[a] |  |  |  |  |
| Single | 34.5 | 49.0 | 71.6 | 47.0 |
| Married | 29.8 | 46.8 | 65.4 | 35.2 |
| Divorced | 35.8 | 49.9 | 68.5 | 40.6 |
| Widowed | 33.5 | 51.6 | 73.1 | 48.8 |
| All | 31.7 | 50.0 | 72.1 | 43.2 |
| N | 53 543 | 37 640 | 12 387 | 103 570 |

[*Source*: 2 per cent Individual SAR for GB]

[a] The percentages for institutions include only non-staff residents. The total percentages include all residents of private households and institutions (including staff).

The higher probability of single rather than currently married women being in institutions, with the formerly married group being intermediate in rank, means that when the total female population is considered, the patterns found for the private household population are modified (Fig. 8.3b), and single women at older ages no longer appear to be the healthiest group. Nevertheless, although the relative advantages

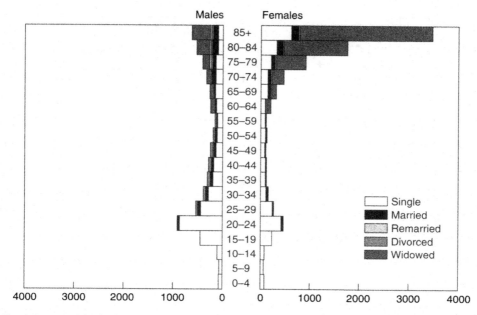

**Fig. 8.2** Non-private household population by marital status, GB 1991. [*Source*: 1991 2 per cent Household SAR]

they enjoy in the private household sector have diminished, even when the much higher proportion of those in institutions is included, single women are recorded as having average levels of health rather than the highest rates found among the younger groups. Thus, if only the private household population is analyzed, the large difference in long-term illness rates between the never married and the married

**Table 8.3** Proportion of adults with a limiting long-term illness aged 65 and over in 1991 resident in institutions (%) by marital status, age and sex

| Age | Marital status | | | |
|---|---|---|---|---|
| | Single | Married/remarried | Divorced/widowed | Total |
| Males | | | | |
| 65–74 | 18.5 | 1.0 | 6.3 | 3.4 |
| 75–84 | 27.4 | 3.6 | 14.6 | 8.5 |
| 85+ | 48.5 | 10.8 | 30.4 | 23.4 |
| ≥65 | 23.5 | 2.4 | 14.5 | 7.1 |
| $N^a$ | 2223 | 18811 | 6914 | 27948 |
| Females | | | | |
| 65–74 | 15.2 | 1.2 | 5.0 | 3.9 |
| 75–84 | 25.9 | 4.4 | 14.7 | 13.1 |
| 85+ | 46.8 | 18.1 | 36.3 | 36.2 |
| ≥65 | 28.3 | 3.2 | 17.8 | 14.2 |
| $N^a$ | 4336 | 14071 | 26278 | 44685 |

[*Source*: 2 per cent Individual SAR for GB]

[a] $N$ is the total number of ill individuals. Resident staff in institutions are not included.

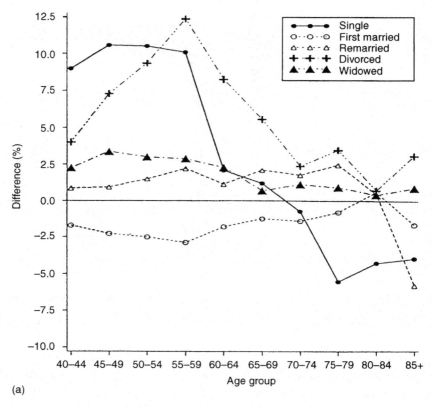

**Fig. 8.3a** Differences in prevalence of long-term sickness in women from average private household population. [*Source*: 1991 2 per cent Individual SAR]

would appear to give the impression that the never-married are better off than the married in the older age groups. Using the total population, married individuals maintain their health advantage, even at older ages.

## 8.2.5 Discussion

The findings reported here raise a number of important questions about the measurement of health status of population subgroups and future trends in the health of the older population as a whole. Firstly, our results reinforce the message that studies of health (and other characteristics) in later life may be seriously biased if the institutionalized population is omitted. Studies which use the private household population to analyze the relationship between health and marital status at older ages show that never-married women at older ages are healthier than married women (Goldman *et al.*, 1995). When the institutionalized population is included, however, the 'conventional' pattern of higher morbidity among the never-married relative to the married is found to prevail. As the proportion of the older population in institutions may vary in

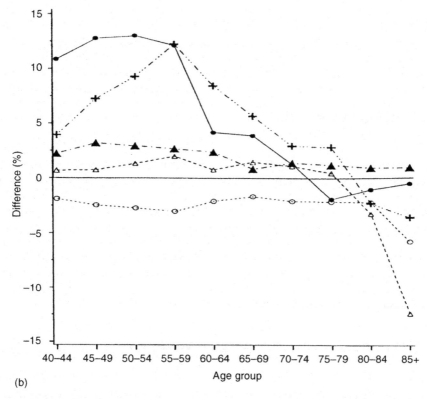

(b)

**Fig. 8.3b** Differences in prevalence of long-term sickness in women from average total population. [*Source*: 1991 2 per cent Individual SAR]

response to policy changes, the extent of this bias may change over time, which makes it particularly difficult to allow for in analyses based on the private household population alone.

## 8.3  A cross-national look at Chinese immigrant occupations
### Suzanne Model

## 8.3.1 Introduction

Until relatively recently, few scholars attempted cross-national studies of immigrants because very little appropriate data were available. Of course, cross-national surveys have existed for many years, but these rarely contain enough minority individuals to study a particular origin group. Censuses can usually support cross-national research because of their near universal coverage. Thus, the commitment by a growing number of governments to, at minimum, the decennial production of census-based microdata samples is a windfall for students of race, ethnicity and immigration. We illustrate these benefits by presenting a multivariate analysis of the socioeconomic status of Chinese immigrants in three nations: Great Britain, Canada and the USA.

The Chinese have already been the object of some cross-national quantitative work. Seeking to test the hypothesis that the USA is an 'exceptionally open' society, Yuan Cheng (1994) compared two outcomes: Goldthorpe class location and unemployment, for Chinese immigrants in the UK and the USA. She drew upon several British Labour Force Surveys and the 1980 PUMS. Logistic regression analysis uncovered no cross-national differences for men or women on either outcome. Model (1997) came to similar conclusions using the 1990 PUMS and 1991 SARs. In addition, Baker and Benjamin (1997a) tested the hypothesis that Canadian society is less discriminatory than that in the USA. Using the 1990 PUMS and 1991 PUMFs, they modelled the earnings of four groups of non-White immigrant males. Their conclusion: only one minority fared better in Canada, the Chinese. The research below adds three insights to this literature: first, Chinese occupational attainment varies not only by where Chinese settle but also by where they were born; second, cross-national differences in the propensity to hold ethnically typed jobs contribute to cross-national differences in Chinese occupational attainment; and, third, controlling for nativity and job type, the few significant cross-national differences that arise tend to favour the USA.

## 8.3.2 Data and deriving comparable variables

The data come from the most recent censuses in each nation: the 1990 US PUMS (5 and 1 per cent), the 1991 Canadian PUMFs (3 per cent), and the 1991 GB SARs (2 and 1 per cent combined). While all three censuses give respondents the opportunity to identify as 'Chinese', many of the variables common to the three sources are coded differently. Moreover, the British data offer fewer variables than the North American data. We discuss only the details involved in addressing the three most challenging data problems: identifying place of birth, creating an occupationally based dependent variable and coding education.

The Chinese are one of several ethnic groups with a relatively long history of emigration. Thus, when admission rules in developed nations warmed to Third World immigrants, Chinese the world over responded. But geography and history created different linkages between Britain, Canada and the USA and the many homelands of the Chinese people. As a result, the countries sending Chinese to these three destinations vary considerably. For instance, slightly over half the Chinese immigrants in the USA and Canada were born in the People's Republic of China, compared with only a third in Britain. Similarly 1 in 5 Chinese immigrants in Canada or Britain was born in Vietnam, versus 1 in 10 in the USA. Finally, only in North America do the Taiwanese comprise a significant component, while only in Britain is Malaysia an important source of Chinese immigrants (Statistics Canada, 1992b; Office of Population Censuses and Surveys, 1993b).

The motivation for noting these diverse origins comes from a recent study of the Chinese in Britain which uncovered significant differences in social class by birthplace (Cheng, 1996). Chinese men born in Hong Kong were significantly less likely than native White men to hold upper white collar jobs, while Chinese born in other nations were more likely to hold such jobs. This result suggests that a cross-national comparison of Chinese socioeconomic status should take birth place into account.

Whilst the USA PUMS code birthplace for all nations of the world, and the Canadian PUMFs identify Hong Kong, the People's Republic of China and Vietnam, less detail is available in the British SARs. The SARs distinguish Hong Kong, Malaysia and Singapore, but group into the 'Other Asia' category the People's Republic of China, Vietnam and Taiwan. Because of these variations in coding, we decided to examine Chinese socioeconomic status at the highest level of refinement possible in each census and to divide each country's residual cases into two country of birth categories: other Asia and non-Asia.

The dependent variable most common to research on the socioeconomic success of immigrants to Britain is social class as defined by Goldthorpe (1987), a measure that rarely appears in the North American literature. Both occupational status and earnings have received considerable attention in the USA, while in Canada most studies examine earnings. This usage reflects not only theoretical preferences but practical realities. The British census does not inquire about earnings and the Canadian census codes occupation very broadly. The research reported here derives an occupational status measure for all three datasets.

The basis for the occupation status measure is the ISEI, an internationally standardized occupational status index (Ganzeboom et al., 1992; Ganzeboom and Treiman, 1996). In the US case, ISEI scores are available only for 1980 occupational titles, but a modest amount of recoding converts 1990 titles into 1980. The SARs poses greater difficulties. The 1 per cent sample already contains this variable because it offers sufficient occupational detail (358 three-digit occupations) to support the addition (see Section 2.5.6). But the 2 per cent sample does not contain the ISEI because its occupational coding is too broad (73 two-digit occupations). Thus, a method for assigning ISEI scores to two-digit occupations was needed. The decision was to aggregate the 358 three-digit occupations into 73 two-digit occupations and calculate the weighted average of the corresponding ISEIs. Each individual in the 2 per cent sample was assigned the weighted average ISEI associated with her or his two-digit occupation. For individuals drawn from the 1 per cent sample, the original value was retained.

The above method provides an introduction to solving the far more vexing problem that the Canadian data provide only 14 occupational codes. Following Ganzeboom et al.'s (1992) observation that 'the costs of being crude' are severe only when the number of occupations falls below six, weighted average ISEIs were computed for each of the 14 Canadian occupations. Calculating these averages required three pieces of information: the detailed occupational codes associated with each of the 14 categories (Statistics Canada, 1994), the title of and number of persons in each detailed occupational code (Statistics Canada, 1995), and the ISEI score for each occupational title (Ganzeboom and Treiman, 1996). To facilitate the mapping process, Ganzeboom and Treiman's 'numerical dominance rule' was used: when a detailed Canadian occupational code contained titles associated with more than one ISEI, the Canadian occupation was assigned the ISEI of its largest title. After each of the 14 occupations had been assigned all its component ISEIs, the ISEIs were weighted by the number of their incumbents. Each individual was then assigned an ISEI equal to the weighted average associated with her or his occupation.

A less time-consuming but equally unfortunate problem arises in conjunction with education. The 1991 British Census distinguishes only three levels of schooling: qualifications obtained after age 18, first degree and higher degrees. No additional information can mitigate this loss; rather, when British outcomes are compared with American and Canadian outcomes, the codes for schooling in the latter two countries must correspond to the British. To attain this objective, we created three levels of post-secondary schooling in the Canadian and US data: diploma/certificate attained after high school but below college, college degree, and postgraduate degree. These credentials, however, are not strictly comparable; indeed, British credentials often have no meaningful counterpart in North America.

## 8.3.3 Analysis and results

In each of the models used, we estimated six equations: one for each gender category in each of the three countries. All equations used ordinary least-square regression (see Ch. 7) to predict occupational status as a function of age, age squared and a set of dummy variables for marital status, education, region and Chinese group membership. In model 1, the Chinese group membership variables were foreign and native-born; in models 2 and 3, we coded birthplace using the most detailed information in each dataset. In all estimates, native-born Whites constituted the excluded category. In addition, model 3 included dummy variables for employment in the restaurant industry and, for women only, in apparel manufacture.

Table 8.4 presents the resulting coefficients. They should be interpreted as the difference in socioeconomic status, net of the control variables, between Chinese and native-born Whites, by birthplace, gender and nation of residence. For example, the socioeconomic status of native-born Chinese men in the USA averaged 2.95 points above that of native-born White men. Coefficients without asterisks are not statistically significantly different from zero.

In model 1, not surprisingly, the native-born Chinese do best. At the same time, foreign-born Chinese men incur no status penalty regardless of nation of residence, but Chinese women suffer significant penalties across nations of residence. This phenomenon may reflect what Baker and Benjamin (1997b) have termed a 'family investment strategy' – an arrangement whereby wives accept dead-end jobs in order to underwrite their husbands' future earnings growth.

Disaggregating the foreign-born by birthplace complicates the picture a good deal. Model 2 shows that Chinese with different birthplaces have different trajectories depending on where they settle. Men and women born in Hong Kong suffer in Britain but not elsewhere; indeed, in Canada and the USA, men born in Hong Kong enjoy higher status than native White men. Another finding is that the Chinese born in Vietnam face the largest shortfalls of any group. This well-documented result is usually attributed to their refugee status. As politically motivated migrants, refugees are less prepared to compete than are economically motivated migrants. Yet, this hypothesis sheds no light on our discovery that the Vietnamese-born have lower status in Canada than in the USA.

Cross-national differences in the extent to which receiving communities channel newcomers into ethnically typed jobs may be one explanation for cross-national

**Table 8.4** Net effects of Chinese nativity on socioeconomic status by gender and nation[a]

| Nativity | Men | | | Women | | |
|---|---|---|---|---|---|---|
| | GB | USA | Canada | GB | USA | Canada |
| *Model 1* | | | | | | |
| NB Chinese | na[b] | 2.95*** | 4.74*** | na | 2.79*** | 1.01 |
| | | (0.37) | (0.62) | | (0.37) | (0.65) |
| FB Chinese | −0.21 | −0.50 | 0.54 | −2.25*** | −2.36*** | −3.71*** |
| | (0.49) | (0.30) | (0.30) | (0.52) | (0.30) | (0.31) |
| *Model 2 with country of birth disaggregated* | | | | | | |
| FB Hong Kong | −1.26* | 2.56*** | 2.44*** | −1.57* | 0.40 | −0.76 |
| | (0.63) | (0.40) | (0.40) | (0.75) | (0.40) | (0.39) |
| FB People's Republic of China | na | −1.64*** | 0.43 | na | −4.15*** | −5.88*** |
| | | (0.31) | (0.38) | | (0.32) | (0.40) |
| FB Vietnam | na | −3.07*** | −4.86*** | na | −3.93*** | −10.29*** |
| | | (0.49) | (0.62) | | (0.50) | (0.62) |
| FB Taiwan | na | 1.18** | na | na | −0.04 | na |
| | | (0.37) | | | (0.35) | |
| FB Malaysia | na | −0.04 | na | na | −2.64** | na |
| | | (0.87) | | | (0.83) | |
| FB Singapore | na | 1.81 | na | na | 0.16 | na |
| | | (1.21) | | | (1.15) | |
| FB Other Asia | 0.45 | 0.71 | 0.27 | −2.25* | −1.93*** | −2.76*** |
| | (0.90) | (0.50) | (0.54) | (1.00) | (0.49) | (0.53) |
| FB Non Asia | 1.84 | −0.07 | 2.24** | −3.02*** | −1.45** | −0.03 |
| | (0.98) | (0.54) | (0.75) | (0.83) | (0.51) | (0.76) |
| *Model 3 with controls for ethnically typed occupations* | | | | | | |
| NB Chinese | na | 4.07*** | 4.99*** | na | 3.63*** | 1.82** |
| | | (0.37) | (0.62) | | (0.36) | (0.60) |
| FB Hong Kong | 1.71 | 3.92*** | 2.49*** | 0.53 | 1.64*** | 0.01 |
| | (0.89) | (0.40) | (0.40) | (0.94) | (0.39) | (0.37) |
| FB People's Republic of China | na | 0.01 | 0.36 | na | −2.20*** | −3.40*** |
| | | (0.33) | (0.40) | | (0.32) | (0.40) |
| FB Vietnam | na | −2.22*** | −4.88*** | na | −3.24*** | −7.13*** |
| | | (0.49) | (0.62) | | (0.49) | (0.60) |
| FB Taiwan | na | 2.42*** | na | na | 0.83* | na |
| | | (0.37) | | | (0.35) | |
| FB Malaysia | na | 1.38 | na | na | −1.59 | na |
| | | (0.86) | | | (0.82) | |
| FB Singapore | na | 2.93* | na | na | 0.87 | na |
| | | (1.20) | | | (1.13) | |
| FB Other Asia | 3.01** | 1.86*** | 0.25 | −0.89 | −0.91 | −1.61** |
| | (1.04) | (0.50) | (0.54) | (1.07) | (0.48) | (0.50) |
| FB Non-Asia | 2.62** | 0.93 | 2.22** | −2.69** | −0.69 | 0.24 |
| | (0.99) | (0.54) | (0.75) | (0.84) | (0.50) | (0.72) |

[a] All models contain the following control variables: age, age squared, and dummies for post-secondary education, marital status and region. Model 3 also contains dummy variables for employment in catering for men and in catering and apparel manufacture for women. The omitted national background is native-born Whites.
[b] 'na' means the data are 'not available', due either to small sample size or to the coding scheme for place of birth.
*** $p < 0.001$; ** $p < 0.01$; * $p < 0.05$.
NB, native-born; FB, foreign-born.

variations in immigrant success. For instance, a higher proportion of Chinese work in catering in Britain than in Canada or the USA. Conversely, North American Chinese women are overrepresented in garment manufacture, but their British counterparts are not. In order to test the relevance of these patterns for socioeconomic status, we estimated a third set of models in which controls were added for these two jobs.

The magnitude of the cross-national differences is generally smaller in model 3 than in model 2, which suggests that cross-national differences in ethnically typed employment contribute to variations in occupational attainment. Another finding is that the remaining differences favour the USA. Tests of statistical significance for each specific sending country (not shown) reveal that all groups of Chinese women hold significantly higher status in the USA than in Canada, while for men a US advantage obtains only for those born in Hong Kong and Vietnam. As for the American–British comparison, Hong Kong born men hold higher status in the USA, but the gap between women is not significant. Canadian–British differences are insignificant.

These results imply that researchers should continue to distinguish Chinese immigrants by birthplace as well as by their propensity to pursue ethnically typed employment. Future work also needs to account for the greater occupational success of the Chinese in the US mainstream economy. Perhaps the small American advantage relative to Britain is due to the 'exceptional openness' of US society, but this explanation sheds no light on the more pervasive American advantage relative to Canada.

## 8.4 Ethnic composition of families in Great Britain and the USA
*Clare Holdsworth*

## 8.4.1 Introduction

The inclusion of a question on ethnic group in the 1991 British Census has proved to be of great value for social scientists and a large amount of research has been based on the information collected (e.g. see Coleman and Salt, 1996). However, there has been relatively little cross-national research to compare the experiences of ethnic groups as identified in the British census with those identified in other national censuses. One reason for this is the problem of cross-national comparability between ethnic group questions, as discussed in Section 3.5. In particular, the relevant questions asked in each country are quite different and refer to distinct conceptualizations of 'race' and 'ethnicity'. In the USA, a question is asked on race and subsequent questions on ancestry (all respondents may identify two (or more) ancestral origins) and Hispanic origin. This differs from the British census, which asks a question on membership of ethnic group. While critics of the British question have argued that in essence the British question only distinguishes *racially* distinct ethnic groups, and thus, in practice, identifies racial as opposed to ethnic origins, there are important differences between the two questions (Leech, 1989; Ní Bhrolcháin, 1990). In particular, in the USA prior to the 2000 Census, the concept of race was dependent on membership of one racial group only. Thus individuals could not identify themselves as 'mixed race', although respondents could give a mixed 'ethnic' origin in the questions for ancestry. Moreover, a separate question was asked on Hispanic origin, which, while not recognized as a *racial* group, was distinguished separately as a quasi-racial/ancestral grouping. In contrast, in the 1991 British census, while there was only one question on 'ethnic group', respondents were able to identify themselves as 'any other ethnic group' and write in their ethnic origin either as a subheading under 'Black' or as a separate group. These two choices for 'any other group' allowed individuals to define membership of more than one group. For most outputs, respondents who gave a mixed ethnic group were coded either as 'Black-other' or 'other ethnic group', although for the minority of tables where 35 ethnic groups were identified, the most common responses to these questions, such as Black/White, are distinguished.

Given the differences in the construction of the US race and British ethnic questions, it is informative to examine the racial/ethnic composition of families to compare the extent of racial/ethnic family homogeneity in the USA and Britain, as identified by these two questions (Dale and Holdsworth, 1997). Given the fact that USA respondents cannot identify themselves as mixed race, it is of particular interest to compare the racial origin of children with parents of different racial origin, with the ethnic origin of British children from mixed-ethnic origin families.

This analysis uses the Household files for the GB and US census. Given the large size of the US household file, I have selected the 1 per cent file for New York to compare with the British 1 per cent Household file. The codings for the questions made available in the microdata files are given below.

## 8.4.2 Coding of questions on race and ethnic group

### Coding for racial group question in US PUMS

| | |
|---|---|
| 001 White | 027 Tahitian |
| 002 Black | 028 Tongan |
| 004 Eskimo | 029 Other Polynesian |
| 005 Aluet | 030 Guamanian |
| 006 Chinese excluding Taiwan | 031 N Mariana Islander |
| 007 Taiwanese | 032 Palauan |
| 008 Filipino | 033 Other Micronesian |
| 009 Japanese | 034 Fijian |
| 010 Asian Indian | 035 Other Melanesian |
| 011 Korean | 036 Pacific Islander not specified |
| 012 Vietnamese | 037 Other Race |
| 013 Cambodian | 301 Alaska/Athabaskan |
| 014 Hmong | 302 Apache |
| 015 Laotian | 303 Blackfoot |
| 016 Thai | 304 Cherokee |
| 017 Bangladeshi | 305 Cheyenee |
| 018 Burmese | 306 Chicksaw |
| 019 Indonesian | 307 Chippewa |
| 020 Malayan | 308 Choctaw |
| 021 Okinawan | 309 Comanche |
| 022 Pakistani | 310 Creek |
| 023 Sri Lankan | 311 Crow |
| 024 All other Asian | 312 Iroquois |
| 025 Hawaiian | 313 Kiowa |
| 026 Samoan | 314 Lumbee |

315 Navajo

316 Osage

317 Paiute

318 Pima

319 Potawatoni

320 Pueblo

321 Seminole

322 Soshone

323 Sioux

324 Tlingit

325 Tohono O Odham

326 American Indian: all other

327 American Indian: not specified

### Coding for ethnic group question in GB SARs

1. White

2. Black Caribbean

3. Black African

4. Black Other

5. Indian

6. Pakistani

7. Bangladeshi

8. Chinese

9. Other Asian

10. Any other group

*For this analysis these have been recoded to broadly equivalent groups as follows:*

### USA (New York)

1. White

2. Black

3. Indian Asian

4. Chinese

5. Other Asian

6. Other group: divided into
   (i)  American Indian
   (ii) Any other group

### Great Britain

1. White

2. Black: divided into
   (i)   Black Caribbean
   (ii)  Black African
   (iii) Black Other

3. South East Asian

4. Chinese

5. Other Asian

6. Other group

## 8.4.3 Inter-ethnic unions

The first part of this analysis compares the proportion of inter-ethnic and inter-racial *marital* unions in the two countries (Tables 8.5 and 8.6). (The steps in using hierarchical data are outlined in Section 5.3.2, with examples of SPSS syntax in Appendix 5.3.)

The main conclusion to be drawn from this analysis is that there are far more inter-ethnic unions in Great Britain than inter-racial unions in New York. In particular, the British data show that just under one-third of married Black men (all three groups combined) and one-fifth of married Black women were married to a partner of a different ethnic group, i.e. with a non-Black wife or husband, usually White. Among the three separate Black groups, this proportion is highest for the Black-Other group. Moreover, a larger number of British Black men were married to White women than vice versa.

In contrast, under 10 per cent of Black New Yorkers are married to someone from a different racial group. It is also interesting to compare the other racial/ethnic group in each country. In Britain, around half of respondents in this group had entered into

Table 8.5 Comparison of racial group of married couples, New York, 1990

| | Partner same racial group | Partner different racial group | | Number of cases |
|---|---|---|---|---|
| | | White | Other racial group | |
| Husband | | | | |
| White | 99.0 | na | 1.0 | 34 171 |
| Black | 92.7 | 5.3 | 1.9 | 2653 |
| Indian Asian | 91.4 | 4.3 | 3.2 | 444 |
| Chinese | 93.7 | 3.2 | 3 | 555 |
| Other Asian | 91.4 | 6.8 | 1.7 | 444 |
| Other group[a] | 84.8 | 11.4 | 3.8 | 1288 |
| (i) American Indian | 42.1 | 50.9 | 7.0 | 114 |
| Wife | | | | |
| White | 99.0 | na | 1.0 | 34 183 |
| Black | 95.3 | 3.0 | 1.7 | 2580 |
| Indian Asian | 96.7 | 2.7 | 0.6 | 332 |
| Chinese | 93.9 | 4.3 | 1.8 | 554 |
| Other Asian | 75.6 | 20.1 | 4.3 | 537 |
| Other group[a] | 85.8 | 9.4 | 4.8 | 1272 |
| (i) American Indian | 42.1 | 43.9 | 14.0 | 114 |

Source: 1990 New York Census of Population, PUMS, 1 per cent sample]
Includes American Indians.

nter-ethnic unions compared with around 20 per cent of New Yorkers. In New York only American Indians enter into a high number of interracial unions. However, Whites and South East Asians recorded a similar proportion of men and women in nter-ethnic unions in both countries, although it should be noted that only a very small minority were married to a partner from a different group.

Table 8.6 Comparison of ethnic group of married couples, Great Britain, 1991

| | Partner same ethnic group | Partner different ethnic group | | Number of cases |
|---|---|---|---|---|
| | | White | Other racial group | |
| Husband | | | | |
| White | 99.5 | na | 0.5 | 115 661 |
| Black | 75.9 | 21.8 | 2.3 | 1042 |
| (i) Caribbean | 75.4 | 19.9 | 4.7 | 678 |
| (ii) African | 74.7 | 16.2 | 9.1 | 253 |
| (iii) Other | 45.9 | 46.8 | 7.3 | 111 |
| South-East Asian | 94.0 | 5.0 | 1.0 | 2942 |
| Chinese | 86.6 | 12.6 | 0.8 | 262 |
| Other Asian | 80.2 | 14.1 | 5.7 | 363 |
| Other group | 47.0 | 47.7 | 5.3 | 394 |
| Wife | | | | |
| White | 99.4 | na | 0.6 | 115 754 |
| Black | 81.8 | 16.3 | 1.9 | 967 |
| (i) Caribbean | 83.1 | 12.8 | 4.1 | 615 |
| (ii) African | 79.1 | 15.1 | 5.8 | 239 |
| (iii) Other | 46.0 | 38.1 | 15.9 | 113 |
| South-East Asian | 96.4 | 2.5 | 1.1 | 2867 |
| Chinese | 73.2 | 23.5 | 3.3 | 310 |
| Other Asian | 65.4 | 30.8 | 3.8 | 445 |
| Other group | 57.6 | 34.6 | 7.8 | 321 |

Source: 1991 British SARs, 1 per cent Household File. Crown Copyright]

## 8.4.4 Children and parents

From the analysis of inter-racial/ethnic unions we would expect there to be more children living with mixed-racial/ethnic parents in Britain than in the USA. In this section we compare the race/ethnicity of children with that of their parents to establish the proportion of 'mixed origin' children and to examine how they are classified by the two census questions. The technique used to generate these tables is given in Chapter 5. We have selected only families with two parents and children (of all ages) resident. These families account for 71.9 per cent of all children in New York and 77.1 per cent of all children in Great Britain (although only 38.0 per cent of Black New York and 48.4 per cent of Black British children).

The results of this analysis are given in Tables 8.7 and 8.8. As expected, a greater proportion of children in Britain live with parents of different ethnic groups, particularly among Black (all three groups, but especially Black-Other) and 'any other ethnic group' children. Among all Black children, the largest number with parents from

**Table 8.7** Comparison of racial group of children with that of parents, New York, 1990

| Racial group of child | Both parents same racial group | Racial group of parents | | | | | |
|---|---|---|---|---|---|---|---|
| | | Mother same group | | Father same group | | Both parents different group | Number of cases |
| | | Father: White | Father: Other | Mother: White | Mother: Other | | |
| White | 98.7 | na | 0.6 | na | 0.5 | 0.2 | 34 539 |
| Black | 92.9 | 1.0 | 0.9 | 3.5 | 0.8 | 0.9 | 2895 |
| Indian Asian | 94.7 | – | 0.6 | 1.6 | 1.0 | 2.2 | 488 |
| Chinese | 93.8 | 2.2 | 0.8 | 1.0 | 1.4 | 0.7 | 762 |
| Other Asian | 79.0 | 5.1 | 1.2 | 0.9 | 0.9 | 13.0 | 488 |
| Other group[a] | 85.2 | 2.4 | 2.4 | 2.7 | 0.6 | 6.6 | 1565 |
| (i) American Indian | 47.5 | 16.1 | 3.4 | 21.2 | 0.8 | 11.0 | 118 |

[*Source*: 1990 New York Census of Population, PUMS. 1 per cent sample]
[a] Includes American Indians.

**Table 8.8** Comparison of ethnic group of children with that of parents, Great Britain, 1991

| Ethnic group of child | Both parents same ethnic group | Ethnic group of parents | | | | | |
|---|---|---|---|---|---|---|---|
| | | Mother same group | | Father same group | | Both parents different group | Number of cases |
| | | Father: White | Father: Other | Mother: White | Mother: Other | | |
| White | 99.3 | na | 0.4 | na | 0.3 | – | 123 139 |
| All Black | 71.4 | 5.5 | 0.4 | 13.0 | 1.1 | 8.6 | 1697 |
| (i) Caribbean | 85.7 | 4.0 | 0.8 | 3.5 | 0.8 | 5.1 | 721 |
| (ii) African | 82.1 | 2.8 | 0.7 | 2.1 | 3.7 | 8.7 | 429 |
| (iii) Other | 14.6 | 3.8 | 3.7 | 7.3 | 2.0 | 68.7 | 547 |
| Indian Asian | 98.1 | 0.2 | 0.2 | 0.5 | 0.4 | 0.5 | 5890 |
| Chinese | 95.9 | 0.3 | 0.5 | – | 0.5 | 2.7 | 368 |
| Other Asian | 79.8 | 3.7 | 0.2 | 1.2 | 2.3 | 12.9 | 519 |
| Other group | 26.7 | 5.8 | 2.6 | 6.7 | 1.8 | 56.4 | 1048 |

[*Source*: 1991 British Census SARs, 1 per cent Household File. Crown Copyright]

different ethnic groups live with a White mother and Black father, reflecting the higher number of inter-ethnic unions of this type. Two-thirds of Black-Other children have both parents from different ethnic groups, of whom a large proportion are children of Black-Caribbean/White unions. This demonstrates that the Black-Other and 'any other ethnic group' categories in the British census include a large number of mixed-ethnic origin children, as they were designed to do. In contrast, a much smaller proportion of New York children live with mixed-race parents, even in the 'other race' group. It is interesting, though, to note that among Black children, the largest proportion with mixed-race parents have a White mother and Black father, the same as in Britain. Moreover, the racial group with the largest number of children with mixed-race parents is American Indian, reflecting the large number of inter-racial unions in this group.

## 8.4.5 Discussion

This analysis suggests that there is a greater degree of ethnic heterogeneity in Britain than racial heterogeneity in the USA. A greater proportion of unions are between individuals from different ethnic groups in Great Britain, particularly between one White and one Black partner and, as a result, more British children live with mixed ethnic parents. However, it should be noted that it is only possible to identify mixed parentage for intact families, which represent a minority of Black children in both New York and GB. Moreover, as suggested in the introduction, any comparison of ethnic/racial groups based on census data from these two countries must take on board differences in the actual questions asked. In particular, it might be the case that as all USA respondents to the 1991 US census had to identify one racial group and could not give a mixed origin as in Britain, they were less likely to record intra-family racial heterogeneity. The classification of racial origin in the USA is a politically sensitive issue for parents and children in mixed-race families who have to choose one racial group for their children. More detailed research on the ways in which British and American ethnic communities record their ethnic/racial identity would be informative for further cross-national research on ethnic and racial issues.

## 8.5 Multilevel modelling of limiting long-term illness using the 1991 Individual SAR for Great Britain
### Myles Gould and Kelvyn Jones

## 8.5.1 Introduction

The richness of data (i.e. the flexibility in the choice of variables and categories) contained in the Individual SAR, together with its hierarchical structure, where individuals are nested within 278 subregional SAR areas in England, Wales and Scotland, has been exploited by health geographers interested in exploring *compositional* (what is in a place) and *contextual* (the difference a place makes) variations in health

(Gould and Jones, 1996; Shouls *et al.*, 1996). Using multilevel modelling we have conducted a detailed analysis of individual SAR records for limiting long-term illness cross-tabulated by sex, age, social class, housing tenure, car ownership and ethnic group (Gould and Jones, 1996). An overview of multilevel modelling is provided in Section 7.4. Section 7.4.3 explains how the data are organized and Section 7.4.4 gives the model details.

## 8.5.2 Models fitted and results

Table 8.9 and Figures 8.4–8.6 provide summaries of the results of analyzing reported limiting long-term illness using two-level models where people are nested within SAR areas (Gould and Jones, 1996). Figure 8.4 is a graphical summary of the estimates of parameters in the fixed part of a model with random intercepts (model 1). This model contains individual-level parameters (the fixed part) and allows health status to vary between places (the random intercept term) but assumes that the effect of geography on health is the same for all individuals. This is a random intercepts model as illustrated in Figure 7.9(b) and described by equation (7.50).

Figure 8.4(a) shows that there is a strong relationship between illness and age, but sex differences are not particularly marked except amongst the older age group. Indeed, illness rates for both males and females aged 30–39 are virtually the same. With respect to ethnicity there is no difference in the probability of being ill for Black and White people in the youngest age category, but differences become increasingly more marked for the older age groups. Variations in morbidity are strongly linked to measures of individual socioeconomic circumstances (Fig. 8.4b). For example, there is an almost fourfold increase in morbidity when comparing White males aged 50–60 who are council tenants in social class IV/V (partly skilled and unskilled) with no car with those who are owner occupiers in social class I/II (professional and managerial) with two or more cars. Inclusion of level 1 fixed effects in model 1 results in a considerable reduction in level 2 variance compared with a null variance components model (results not shown here) and provides some evidence of a compositional explanation for geographical variations in long-term limiting illness.

However, the estimated level 2 SAR area variance is 0.07 and is many times its standard error (Table 8.9) and therefore suggests that significant contextual differences between SAR areas remain. These geographical variations in self-reported morbidity

**Table 8.9** Estimates of random terms

|  | Model 1 | Model 2 |
|---|---|---|
| Level 2: SAR areas | | |
| Between-area variance: males (base category) | 0.07 (0.01) | 0.09 (0.01) |
| Between-area variance: female contrast | | 0.01 (0.003) |
| Covariance: base category and female contrast | | −0.02 (0.005) |
| Level 1: Individuals | | |
| Between cell variance | 1.04 (0.01) | 1.04 (0.01) |

[*Source*: Models B and C in Gould and Jones (1996). Reproduced with permission]
Note: standard errors are shown in parentheses.

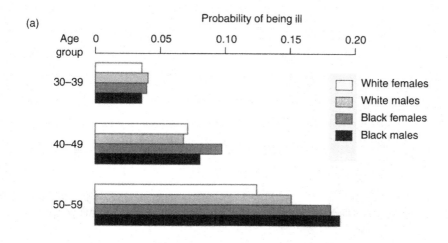

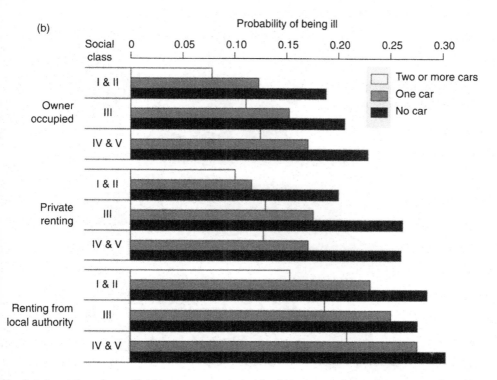

**Fig. 8.4** Proportion who are ill: (a) by age, sex, and ethnicity; (b) by tenure, class and access to car for White males who are aged 50–60. (From model B, Gould and Jones, 1996; reproduced with permission from Elsevier Science.)

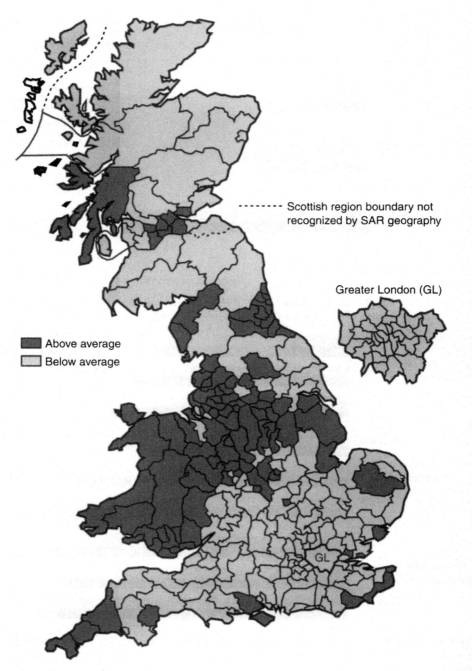

**Fig. 8.5** Variations in limiting long-term illness between SAR areas. (From model B, Gould and Jones, 1996. Boundary map reproduced from OPCS 1992, *Population Trends* **69**, Office for National Statistics, © Crown copyright 1999, with permission.)

ιre mapped in Fig. 8.5 (using the level 2 residuals from the random intercepts model). The pattern is of low morbidity in South-East England (including London), South-West England and rural areas of the North of England and Scotland, whilst there is ιigh morbidity in Wales, the urban North-East and North-West England, and Strath-lyde. The differences are geographically marked, with high levels of illness being found ιn the north and west of the country, having allowed for a large number of sociodemo-ιraphic variables. It is important to emphasize here that this geographical variation is ιhe 'contextual' effect that remains after allowing for individual-level characteristics ('compositional' effects). This is in contrast to Sloggett and Joshi's (1994) findings based on an analysis of the Longitudinal Survey) that excess premature mortality in ιeprived areas is 'wholly explained' by individual characteristics.

Model 2 (Table 8.9) extends model 1 by allowing for differential (random) vari-bility for women at level 2. That is, it tests whether the relationship between ιealth and place is different for women than for men or, in other words, allows for ι different geography of health for men and women. This is equivalent to the ιnodel illustrated in Figure 7.9(c) and described by equation (7.51).

The level 2 variance around the base category for males is 0.09, while for females it ι 0.06, this differential being clearly illustrated in Figure 8.6 by variations in the log-ιdds of being ill (with equivalent probabilities) for all 278 SAR areas. There are

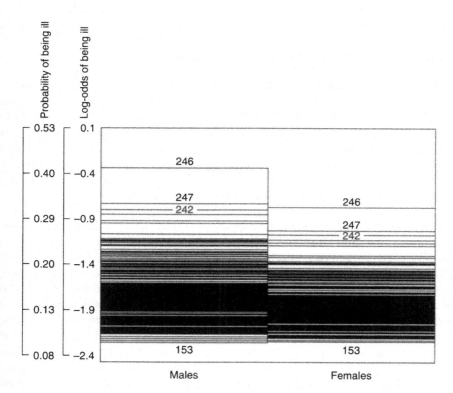

**g. 8.6** Between-SAR area variations in limiting long-term illness for males and females. (From model C, ιould and Jones, 1996, reproduced with permission from Elsevier Science.)

significant place differences in morbidity for men, and significantly different contrasts for women. Places that are high for men are also high for women, but the differences are less marked for the latter group. Indeed, in no part of the country is the female rate of illness higher than that for men. There is a fourfold variation in morbidity rates between the lowest and highest SAR areas for males compared with a threefold variation for females.

Additional analysis involved the fitting of other models where the age and social class categories were also allowed to vary at level 2. These further results demonstrated that there are no significantly different geographical variations in reported illness between older and younger people, or for people in different social classes. In other words, there was no need for different 'maps' of illness for young, middle-aged and old people, or for people in different social classes. The geography of morbidity for these models is well summarized by the overall pattern shown in Figure 8.5.

## 8.5.3 Extending the models

These examples of multilevel modelling demonstrate the value in having flexibility in the choice of SAR variables and categories; and also the insights into sociological and geographical variations in (ill) health which can be uncovered when using a multilevel approach to simultaneously explore individual and contextual relationships. The analysis summarized above could be extended and developed in a number of ways. The analyses could be undertaken on specific population groups, such as Asians or the unemployed, both of which are made possible by the SARs large sample size (see also Fieldhouse and Gould, 1998).

Model 2 could also be extended to include level 2 explanatory variables, i.e. variables measured at the area level that may help in explaining variations in health by SAR area. These may be ecological variables such as environmental pollution or 100 per cent tabular data from the census aggregated to SAR areas (e.g. SAR area unemployment rates or percentage ethnic minorities in a SAR area). Aggregate 100 per cent census data for SAR areas are available from MIMAS (Manchester Information and Associated Services, http://man.ac.uk/mcc/mimas/census/).

Ecological data can also be incorporated into multilevel models using *cross-level interactions* between individual and ecological characteristics with a range of different formulations (Jones and Duncan, 1995). A selection of possible scenarios is shown in Figure 8.7. The vertical axis on each graph represents the individual response (probability of being ill), the horizontal axis the ecological variable (per cent unemployed), while the solid and dotted lines represent individuals of high and low social class, respectively. Figure 8.7(a) shows that there are marked differences between types of individuals but no ecological effect, while Figure 8.7(b) represents the converse: little difference between types of people but a substantial ecological effect. The parallel lines of Figures 8.7(c) and (d) represent the cases when both the individual and ecological effects are marked; in the case of the former, the individual and ecological effects operate in the same direction (people of low social class living in high unemployment areas are less healthy), whilst in the latter case the individual and ecological effects operate in the opposite direction. Figure 8.7(e) represents a model

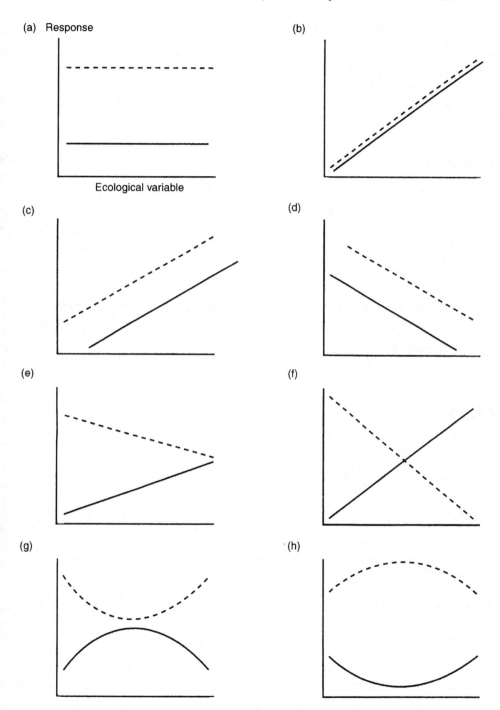

**Fig. 8.7** Individual and ecological cross-level relationships. (From Jones and Duncan, 1995; reproduced with permission from Elsevier Science.)

in which the area effect for unemployment is positive for people of high social class but negative for the working class, while Figure 8.7(f) represents the case where the cross-level interactions are strong enough to invert the individual-level effects. Figures 8.7(g) and (h) portray models in which the non-linear interaction terms are of importance so that either the smallest or largest ecological effects are found for 'middling' levels of unemployment. It is only with the development of the multilevel model that the full complexities of such relationships can be effectively analyzed.

## 8.5.4 Acknowledgements

This section summarizes analysis previously published in Gould and Jones (1996) (copyright Pergamon Press). The SARs were provided through the Census Microdata Unit of the University of Manchester, with the support of the ESCR/JISC/DENI. All tables containing SAR data, and the results of analysis, are reproduced with the permission of the Controller of Her Majesty's Stationery Office (Crown Copyright).

## 8.6 Deprived people or deprived places? Exploring the ecological fallacy in studies of deprivation using the Samples of Anonymised Records
### Ed Fieldhouse[2]

## 8.6.1 Introduction

Area-level deprivation scores are frequently used as a basis for formulating public policy, in particular for targeting resources within local authority areas (Simpson, 1993) and in selecting local authority areas for urban aid. For example, the Department of Environment (DoE, 1983) used six census variables to produce a single index of deprivation which was subsequently used as a basis for discussion over the allocation of resources. This type of approach provides a convenient way of allocating resources between areas and reflects the assumption that deprivation is, in part, spatially determined and cannot be tackled solely by individually oriented policies (such as welfare benefits) (Kirby, 1981).

However, this assumption may be controversial. Any area-based analysis is subject to the danger of ecological fallacy, or the spurious inference of individual-level relationships from area-level data (Robinson, 1950). In other words, just because people live in 'deprived areas' does not necessarily mean they are deprived (Hamnett, 1979). To make this assumption would require social homogeneity of geographical units. Deprivation can take a variety of forms, and therefore different areas may contain different forms of deprivation; whilst some may be spatially concentrated, others may be widely distributed (Hatch and Sherrot, 1973; Cullingford and

[2] Based on an earlier version published in *Environment and Planning A* **28**, 237–59, 1996.

Openshaw, 1979). Consequently, not only will different areas contain different mixes of deprivation, but as Holterman (1975, p. 44) observed:

> on single indicators the spatial concentration of individual aspects of urban deprivation is relatively low, so that priority area policies aimed at single problems would leave many of the deprived outside and therefore raise serious questions of equity.

Similarly, policies targeted on the basis of an area-level index of deprivation are likely to provide no assistance to deprived people living beyond the boundaries of 'deprived areas'.

However, Kirby (1981) has argued that census-based analysis is only capable of identifying 'deprivation which stems from wider social inequalities' rather than that which arises from localized factors such as health care or social services and that deprivation is 'an issue of consumption, locally fostered and consequently only spatially overcome'. This highlights what Smith (1977) has referred to as *place* deprivation or deprivation which is a product of locationally specific access to goods and services. However, Smith also identifies *person* deprivation, which is a product of an individual's position in the broader economic system. In order to sustain the argument that deprivation is spatially determined and best dealt with by area-based policies, it is necessary to demonstrate that person deprivation is spatially concentrated, not diffused.

The analyses presented in this section aim to examine the geographical distribution of deprived *people* and explore the extent to which they are concentrated in deprived *areas*. Rather than concentrating on any one aspect of deprivation, we shall focus on individuals who suffer from 'multiple deprivation'. Using census microdata, it is possible to measure deprivation at the individual level. In doing so, we are able to directly examine the levels of deprivation amongst particular groups and, perhaps more importantly, to go some way towards resolving this question of spatial concentration.

## 8.6.2 Measuring micro-level deprivation with the SARs

Deprivation is a complex composite concept and cannot be represented by a single indicator. However, as with other census data, the selection of variables useful for measuring deprivation in the SARs is limited (Hirschfield, 1994). For example, it is not possible to look at social deprivation or more sophisticated aspects of material deprivation such as diet or possession of consumer goods. Furthermore, whereas area-level analyses have frequently used variables measuring the percentage belonging to an at risk group (e.g. ethnic minorities or the elderly), the indicators used here are restricted to direct material-based measures of relative deprivation. Whilst the use of indirect measures may provide predictive power with respect to some specific outcome (e.g. demand for health care), their inclusion can lead to problems of circularity. For example, indexes using the percentage belonging to ethnic minorities or lone parents will tend to identify areas with a high proportion of these groups as being deprived. Tautologically this may lead to spurious claims that those groups are themselves deprived. It is thus important to distinguish between direct measures of (actual) deprivation and indirect measures *associated* with deprivation (DoE, 1992). Similarly, at the individual level, rather than building ethnicity into the definition of deprivation, it becomes possible to examine the different levels of deprivation in different ethnic groups.

The variables were selected from the 2 per cent Individual SAR and were chosen to measure basic material items or resources, the lack of which may put those individuals at a relative material disadvantage as compared with the rest of the population. The items were organized into three distinct groups using principal components analysis. A multiply deprived person was defined as any person without at least one item from two of the three groupings. The three groupings were:

- Group 1
  No bath
  No inside w.c.
  Not self contained accommodation
- Group 2
  No cars
  No earners
  Tenure (rented)
  No central heating
- Group 3
  >1 person per room
  Unemployment.

Under this definition of deprivation, 8.6 per cent of the population of Great Britain are classified as deprived. Although the threshold is necessarily arbitrary, intuitively this seems a reasonable cut-off point (see Mack and Lansley, 1985).

## 8.6.3 Who's deprived?

Because deprived individuals as opposed to deprived areas have been identified, it is possible to investigate the sociodemographic characteristics of the deprived population without making potentially spurious inferences from area data. In other words, rather than inferring that a particular group is deprived simply because it lives in a deprived area, it is possible to measure this directly. Other studies of deprivation and inequality have indicated sociodemographic differences in the propensity to deprivation (e.g. Berthoud, 1976). There are at least five main cleavages identifiable in the 2 per cent SAR which were shown to be important: social class, ethnicity, age, family type and geography. The remainder of this article focuses on the geographical distribution. Fuller information about the other dimensions can be found in Fieldhouse and Tye (1996).

## 8.6.4 Deprived areas and deprived people: the geographical coincidence of individual- and area-level deprivation

Many of the individual components of deprivation under analysis have distinct geographical patterns (e.g. unemployment) as do the populations most prone to

Table 8.10  Deprivation by SAR Area (25 most deprived areas)

| SAR area | Deprived (%) | Cumulative population (%) | Cumulative deprivation (%) |
|---|---|---|---|
| Tower Hamlets | 35.6 | 0.3 | 1.1 |
| Hackney | 26.0 | 0.6 | 2.1 |
| Newham | 22.2 | 1.0 | 3.0 |
| Islington | 20.7 | 1.2 | 3.7 |
| Camden | 20.1 | 1.5 | 4.4 |
| Southwark | 19.9 | 1.9 | 5.2 |
| Glasgow | 19.6 | 3.1 | 8.0 |
| Lambeth | 19.6 | 3.5 | 8.9 |
| Haringey | 19.6 | 3.8 | 9.6 |
| Motherwell | 18.2 | 4.1 | 10.2 |
| City of London/Westminster | 17.8 | 4.4 | 10.7 |
| Hammersmith/Fulham | 17.2 | 4.6 | 11.2 |
| Knowsley | 16.4 | 4.9 | 11.8 |
| Kensington/Chelsea | 16.3 | 5.1 | 12.2 |
| Liverpool | 16.1 | 6.0 | 13.7 |
| Birmingham | 16.0 | 7.7 | 16.9 |
| Brent | 15.5 | 8.1 | 17.7 |
| Blackburn | 15.5 | 8.4 | 18.1 |
| Manchester | 15.4 | 9.1 | 19.4 |
| Cumbernauld etc. | 14.8 | 9.4 | 19.9 |
| Lewisham | 14.5 | 9.8 | 20.6 |
| Bradford | 14.3 | 10.6 | 22.0 |
| Waltham Forest | 14.3 | 11.0 | 22.6 |
| Wandsworth | 13.8 | 11.4 | 23.3 |
| Kingston/Hull | 13.5 | 11.9 | 24.1 |

Source: 2% Individual SAR and LBS. Crown Copyright]

deprivation (Forrest and Gordon, 1993). These patterns reflect the complex social and economic geography of Great Britain and geographically specific responses to those differences (Lewis and Townsend, 1989). Consequently, it would be expected that multiple deprivation should itself be spatially concentrated, as other studies have shown (e.g. Holterman, 1975; Begg and Eversley, 1986; Bradford et al., 1993; Townsend et al., 1988).

Table 8.10 shows the 25 SAR areas containing the highest percentage of deprived people. One important quality of this measure of deprivation at an area level is that it naturally weights the data towards those variables which contribute most to multiple deprivation. High rates of multiple deprivation are found, not surprisingly, in metropolitan Britain, particularly in Inner London. Tower Hamlets has the highest rate at 35.6 per cent, followed by four other London boroughs. Because areas are ranked in terms of percentage deprived, there may be some bias against larger SAR areas where the deprived population will be more diluted. This may partially explain the domination of London boroughs of the top six, which tend to have a smaller population than other metropolitan areas. The basic message, however, is quite simple: deprivation (as it is defined here) is an urban problem.

Figure 8.8 shows that a large number of areas contain between 4 and 10 per cent deprived persons and only a few contain percentages in excess of this. Consequently, although a few areas appear to contain abnormally high proportions of deprived persons (reflected in the positive skew in the figure), most areas contain a substantial number of deprived people. The degree of concentration is also evident in Table 8.10,

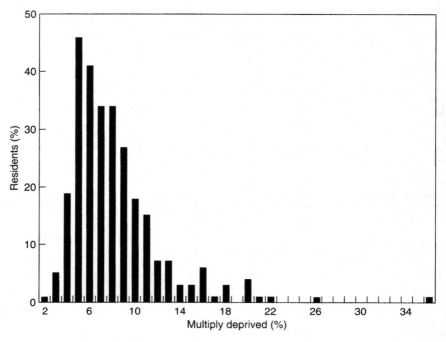

**Fig. 8.8** Percentage of multiply deprived residents by SAR area. SD = 4.08; mean = 8.2; $N$ = 278.00. (Reproduced with permission from Fieldhouse and Tye, 1996.)

which shows the proportion of the British total found in the 25 areas containing the highest proportions of deprived people. This is contrasted with the share of the total population in those areas.

Table 8.10 indicates that, although there is a concentration of deprived people in the most deprived areas, the most striking feature of the distribution of the deprived population is its spread. Roughly one-fifth of Britain's multiply deprived people live in areas containing one-tenth of the total population. Thus, although there is a clear hierarchy of areas identifiable as deprived, the most deprived areas do not contain a majority of the deprived population. Policies aimed at selectively targeting deprived areas are therefore likely to miss a large proportion of deprived people. The story is similar when looking at the social and demographic dimension of deprivation. That is, there are substantial concentrations in some areas (or groups), but these by no means make up the majority of the total deprived population.

## 8.6.5 Conclusions

The focus of this section has been the implications of using individual as opposed to area-level data in the study of deprivation. The use of individual-level data from the SARs has demonstrated that, although multiple deprivation is heavily concentrated in particular social groups and in particular geographical areas, these do not make up the majority of the deprived population in the country.

## 8.7 Small area estimation using census microdata in the UK

*Stephen Ludi Simpson*

### 8.7.1 Introduction

Small area estimation is distinctive from analysis of national-level data (Sections 8.2–8.4) and from the impact of place on individual-level characteristics (Section 8.5). Rather, its focus is the characteristics of a local area, usually in order to provide the information needed to plan services or meet customer demand. Thus small area estimation is used by councils, utilities, health service providers, market research agencies and other organizations, but always with a focus on specific geographical areas such as towns or other communities.

In Britain, standard census output provides tabulations for local areas using 100 per cent of the census records (the Small Area Statistics and Local Base Statistics from the 1991 Census). These give direct counts for small areas. Census microdata can help fill in the gaps in these small area tabulations, and can also help to estimate for local area information (such as income) which cannot be obtained from the census.

A simple example will set the scene. The number of young people living alone in a local authority district is not available from the 100 per cent census tabulations. However, an estimate of the correct value can be extracted from sample microdata. As a second step, expenditure patterns – which are not recorded in the census – of young people living alone in the local area can be inferred using non-census data from regional or national surveys. This section reviews the variety of ways in which sample microdata from the 1991 UK Census may be used for small area estimation – despite the fact that the geographical definition available in microdata samples will often be at a coarser level than that required. These methods improve the reliability of the direct estimate from microdata by combining it with more reliable national data and with relevant (although partial) data from 100 per cent census tabulations. Characteristics which are not recorded by the census can be inferred for small areas by knowledge of their relationship with census variables. In addition, aggregate data, including non-census data, can be adjusted to avoid the ecological fallacy by using microdata to establish the basis for correlation between individuals or households. Finally, subsamples of microdata can be used to simulate areas smaller than those to which the microdata are coded.

### 8.7.2 Supplementing a small sample

At the finest geographical definition (120 000 population from 1991 SARs), samples of microdata are often very small. For example, economic activity rates by age and sex for each ethnic group are required for labour force forecasts in multi-ethnic areas. These are not available from 100 per cent tabulations, while microdata samples contain less than 10 residents for many age groups. To improve the reliability of these estimates, Bradford Council (1996) supplemented SAR data for Bradford with SAR data from larger regions containing Bradford – West Yorkshire, England and Wales – until the sample reached a minimum of 100 for a given cell.

It is common practice to use national data to replace or improve the reliability of a small sample for a single area. Longford (1998, 1999) removes the arbitrary element in the above procedure. The local sample is given a weight relative to the national sample according to (a) the size of the local sample and (b) the variance between all local areas in Britain. If the latter is large, it is taken to indicate that a local divergence from the national mean is more likely to be real than to be a sampling error. He proves that this is the most accurate linear combination of local and national sample proportions and demonstrates the improvement in accuracy over the direct estimator (the local sample alone) for economic activity. His univariate methods can be implemented on a spreadsheet and are precisely the same as that used by specialist but less accessible multilevel statistical software such as MLWin to calculate 'shrunken expected values' (Goldstein, 1995; Goldstein et al., 1998).

### 8.7.3 Consistency with 100 per cent tabulations using iterative proportional fitting

Microdata from the 1991 SARs give results which are inconsistent with the less detailed 100 per cent local tabulations. Continuing with the example of economic activity by each age, sex and ethnic group, the number of residents in each age category obtained from the sample data will give, when scaled by the appropriate sampling fraction, numbers which are different from the 100 per cent count available for age. In part, this inconsistency is due to sampling error; other, lesser, inconsistencies arise because the SARs do not contain imputed records for households missed by the census (Section 4.3). The inconsistency is likely to be largest with smallest samples. To gain consistency and, by assumption, greater accuracy – the microdata results can be scaled to be consistent with the coarser 100 per cent tabulation using iterative proportional fitting (IPF), as for example in Bradford Council (1996) to develop forecasts of the labour force.

IPF can also be used to apply a pattern from the microdata to the known characteristics of areas smaller than those coded in the microdata. Simpson and Middleton (1998) estimate detailed tabulations for each ward of England using IPF with microdata for large areas and 100 per cent tabulations for wards.

In general, IPF requires initial values of a tabulation of two or more variables, which in this case are provided by the microdata. These values are scaled successively to be consistent with the marginal totals of the same variables from a more reliable but less detailed source. Rees (1994) gives a step-by-step description of IPF in another setting, while Bishop et al. (1975) prove that the procedure maintains the interactions of the original microdata and also describe its historical origins and statistical properties. IPF is one method of synthetic estimation, discussed more generally in the following section.

## 8.7.4 Synthetic estimation of non-census data

Synthetic estimation is a long-standing tool of small area estimation (Ghosh and Rao, 1994). In general, it takes relationships from a national (or large area) sample survey and applies them to local areas' census data. The link between the two data sources is made using variables common to both. Estimates of non-census characteristics in the local area are derived on the assumption that the national relationships also hold locally.

Synthetic estimation is often carried out with 100 per cent local census tabulations, e.g. when applying national disability rates by age and sex to a local age–sex structure. Charlton (1998) provides synthetic estimates of local rates of serious illness in which the use of census microdata is essential in two ways. Serious illness is first predicted from a national survey of consultation with general practice doctors, using a logistic regression model with predictors that are also measured in the census – age, sex, employment position, tenure, social class and family type. Because the regression model is non-linear, it cannot be used with aggregate 100 per cent tabulations to predict local levels of serious illness. The microdata also have to be used because the model is different for each age–sex category, and the predictors are not all tabulated with age and sex in the 100 per cent tabulations. Charlton applies the model to each individual in the 1991 Individual SAR and thus calculates the probability that each will have serious illness; when added, these provide an estimate of the total number with a serious illness in each of the SAR areas.

The same approach of synthetic estimation with census microdata has been used to add mean earnings information to each microdata record (Section 5.2.4). It has also been used to enhance census data to include an allowance for non-response (Simpson and Middleton, 1999).

## 8.7.5 Removing aggregation bias in ecological correlation

Occasionally there is a need to estimate the correlation for individuals in a particular area when only aggregated data are available for each variable (i.e. the variables are not cross-tabulated within the area). The data may come from non-census sources, e.g. morbidity data from doctors for each enumeration district, or flows off the unemployment register for each electoral ward, with a need to correlate this with other characteristics for which small area census data are available.

Because local areas are homogeneous, a correlation calculated from aggregated data is likely to be a highly inflated representation of the true correlation between individuals. Tranmer and Steel (1998) investigate four regions – three microdata regions in England and one in Australia – and find that the 'grouping variables' most associated with homogeneity of enumeration districts are similar in each region: age structure, housing type and ethnicity. They then show how knowledge of how the grouping variables behave – using microdata – allows an adjustment for the correlation between other variables based on aggregated data. They demonstrate that this adjustment greatly reduces the aggregation bias. They have thus found a way to avoid the ecological fallacy. The suitability and acceptability in practice of this method will be tested in their future work.

## 8.7.6 Micro-simulation of whole populations

The work described so far has used microdata to provide specific new tabulations and associations for a small area. Micro-simulation attempts to assemble complete individual census records for the smallest areas – enumeration districts in England and Wales. Any new tabulation can then be derived with little further work. For an enumeration district of *n* households, the method seeks a subsample of *n* records from the household microdata that best matches the 100 per cent and 10 per cent tabulations for the enumeration district.

In practice, microdata subsamples do not recreate the enumeration district exactly. The 2 per cent Individual SAR, although containing over 1 million records in total, will not necessarily contain the diversity found in an enumeration district of 250 households. In addition, the differences between microdata and local tabulations with respect to coverage and data modification make an exact fit unlikely. Not surprisingly, atypical enumeration districts are more difficult to simulate than 'average' areas, whose households will be similar to many represented in the national SARs.

Williamson *et al.* (1998) describe and compare algorithms for searching the SARs for subsamples that match chosen local tabulations. They aim to reduce the discrepancy between the local tabulations and a random subsample of the microdata. The best methods reduce that discrepancy by around 70 per cent, thus recreating a considerable amount of the diversity between enumeration districts – although not all of it. Williamson *et al.* (1998) validate the simulation by recreating tabulations that were not used in the simulation itself and find good results. This supports the conclusion of Tranmer and Steel (1998) that only a few variables account for most of the homogeneity of local areas.

The methods of micro-simulation are highly computer-intensive; they are not for the casual user. However, the potential of micro-simulation should not be underestimated, since if successful, the resulting database of individual records could be used by others with little extra outlay.

## 8.7.7 Discussion

Methods of small area estimation attempt to create local statistics which were not included in conventional census output, either because of lack of resources or because of the need to protect the confidentiality of individuals.

An increase in the availability of 100 per cent data for local areas from the 2001 Census may provide direct estimates of census characteristics that currently require the methods of small area estimation described here. Easy access to the full census database for tailor-made tabulations after the standard tabulations have been published may be the most significant development in the next 10 years. Larger samples of microdata for subsets of census variables may also help.

However, there will always be a limit imposed by the need to protect the identity of the real people 'behind' individual census records. Some small area estimation can be seen as an attempt to overcome the confidentiality restrictions on the census – and the micro-simulation of small areas most obviously so – without threatening privacy of individuals.

Small area estimates are inexact, to an extent that users of 100 per cent census tabulations may not be accustomed. Developments to validate the accuracy of each method, represented in much of the work described here, are a parallel and necessary complement to the dissemination of the methods themselves. Methods of small area estimation have been stimulated by the existence of microdata in the UK. Those reviewed here are not exhaustive and some of the work on the topic is carried out by local organizations without publication.

# References

Albert, L., Austin, D. and Boyko, E. 1996: The benefits of census public use microdata files: a Canadian tradition. *SARs Newsletter* No. 7, 9–15.

Al-Hamad, A., Hayes, L. and Flowerdew, R. 1997: Migration of the elderly to join existing households: evidence from the Household SAR. *Environment & Planning A* **29**(7), 1243–1255.

Anthias, F. and Yuval-Davis, N. 1992: *Racialized boundaries: race, nation, gender, colour and class and the anti-racist struggle*, London: Routledge.

Australian Bureau of Statistics 1991: *1991 census dictionary* (cat. no. 2901.0). Canberra: ABS.

Australian Bureau of Statistics 1993: 1991 PES: comparison of census and PES responses. *Census Working Paper 93/4*. Canberra: ABS.

Australian Bureau of Statistics, 1995: *Census 91: data quality – undercount* (cat. no. 2940.0). Canberra: ABS.

Australian Bureau of Statistics 1996: *How Australia takes a census* (cat. no. 2903.0). Canberra: ABS.

Baker, M. and Benjamin, D. 1997a: Ethnicity, foreign birth and earnings: a Canada/US comparison. In Abbott, M.G., Beach, C.M. and Chaykowski, R.P. (eds), *Transition and structural change in the North American labour market*. Kingston, ON: John Deutsch Institute and Industrial Relations Centre, Queen's University, 281–313.

Baker, M. and Benjamin, D. 1997b: The role of the family in immigrants' labor market activity: an evaluation of alternative explanations. *American Economic Review* **87**, 705–727.

Ballard, R. and Kalra, V.S. 1994: *The ethnic dimensions of the 1991 census: a preliminary report.* Manchester Census Group Working Papers. Manchester: Centre for Census and Survey Research.

Barnett, V. 1991: *Sample survey principles and methods*. London: Edward Arnold.

Barr, R. 1993: Mapping and spatial analysis. In Dale, A. and Marsh, C. (eds), *The 1991 census user's guide*. London: HMSO, ch. 9.

Begeot, F., Smith, L. and Pearce, D. 1993: First results from Western European censuses. *Population Trends* **74**, 18–23.

Begg, I. and Eversley, D. 1986: Deprivation in the inner city: social indicators from the 1981 census. In Hausner, V. (ed.), *Critical issues in urban economic development*, vol. 1. Oxford: Clarendon.

Bellente, D. and Kogut, C.A. 1998: Language ability, US labor market experience and the earnings of immigrants. *International Journal of Manpower* **19**(5), 319–332.

Berk, R. A. 1983: An introduction to sample selection bias in sociological data. *American Sociological Review* **48**, 386–398.

Berthoud, R. 1976: *The disadvantages of inequality*. London: MacDonald and Janes.

Bethlehem, J.G., Keller, W.J., Pannekoek, J. 1990: Disclosure control of microdata. *Journal of the American Statistical Association* **85**, 38–45.

Birch, M. and Elias, P. 1997: *ISCO 88 (COM) – a common classification of occupations for Europe.* Warwick: IER, University of Warwick.

Bishop, Y., Fienberg, S. and Holland, P. 1975: *Discrete multivariate analysis.* Massachusetts: MIT Press.

Blackburn, R. M., Dale, A. and Jarman, J. 1997: Ethnic differences in attainment in education, occupation and lifestyle. In Karn, V. (ed.), *Employment, education and housing among ethnic minorities in Britain.* London: HMSO.

Blackburn, R., Siltanen, J. and Jarman, J. 1995: The measurement of occupational gender segregation: current problems and a new approach, *Journal of the Royal Statistical Society, Series A* **158**, 319–331.

Bone, M.R., Bebbington, A.C., Jagger, C., Morgan, K. and Nicolaas, G. 1996: *Health expectancy and its issues.* London: HMSO.

Börsch-Supan, A.H. 1990: A dynamic analysis of household dissolution and living arrangement transitions by elderly Americans. In Wise, D.A. (ed.), *Issues in the economics of aging.* Chicago: National Bureau of Economic Research.

Boyle, P.J. 1995: Public housing as a barrier to long-distance migration. *International Journal of Population Geography* **1**, 147–164.

Boyle, P.J., Cooke, T., Halfacree, K.H. and Smith, D. 1999a: Integrating GB and US census microdata for studying the impact of family migration on partnered women's labour market status. *International Journal of Population Geography* **5**, 157–178.

Boyle, P.J., Cooke, T., Halfacree, K.H. and Smith, D. 2000: Rethinking long-distance family migration: the effect on partnered women's employment status in GB and the US. Personal communication, mimeo, University of St. Andrews, Scotland.

Boyle, P.J. and Halfacree, K.H. 1996: Gender issues in the migration of single, service class adults in Britain: an urban perspective. *RGS/IBG Annual Conference*, Strathclyde University, Glasgow, 3–6 January.

Boyle, P. and Halfacree, K.H. (eds), 1999: *Migration and gender in the developed world.* London: Longman.

Boyle, P.J., Halfacree, K.H. and Smith, D. 1999b: Family migration and female participation in the labour market: moving beyond individual-level analyses. In Boyle, P.J. and Halfacree, K.H. (eds), *Migration and gender in developed countries.* London: Routledge, 114–135.

Bradford Council 1996: *Forecasts of the labour force: technical report.* Bradford: Policy and Research Unit, Corporate Services, City Hall.

Bradford, M.G., Robson, B.T. and Tye, R. 1993: Constructing the 1991 urban deprivation index. *Spatial Policy Unit working paper 24.* Manchester: School of Geography, University of Manchester.

Breen, R. 1996: *Regression models: censored, sample selected and truncated.* Quantitative Applications in the Social Sciences. London: Sage.

Breen, R. and Hayes, B.C. 1996: Religious mobility in the UK. *Journal of the Royal Statistical Society, Series A* **159**(Part 3), 493–504.

Britton, M. and Birch, F. 1985: *1981 census post-enumeration survey: an enquiry into the coverage and quality of the 1981 census in England and Wales.* London: HMSO.

Brown, A. 1978: Towards a world census, *Population Trends* **14**, 17–19.

Brown, J.J., Diamond, I.D., Chambers, R.L., Buckner, L.J. and Teague, A. 1999: A methodological strategy for a one-number census, *Journal of the Royal Statistical Society, Series A* **162**, 247–267.

Bruegel, I. 1999: Who gets on the escalator? Migration, social mobility and gender in Britain. In Boyle, P.J. and Halfacree, K.H. (eds), *Migration and gender in the developed world.* London: Longman.

Bryman, A. and Cramer, D. 1990: *Quantitative data analysis for social scientists*. London: Routledge.

Buck, N., Gershuny, J., Rose, D. and Scott, J. 1994: *Changing households, the British Household Panel Study, 1990–1992*. Essex: Centre for Micro-Social Change.

Bullen, N., Jones, K. and Duncan, C. 1997: Modelling complexity: analysing between-individual and between-place variation – a multilevel tutorial. *Environment and Planning A* **29**, 585–610.

Bulmer, M. 1986: A controversial census topic. *Journal of Official Statistics* **2**, 471–480.

Burr, J.A. 1990. Race/sex comparisons of elderly living arrangements. *Research on Aging* **12**(4), 507–530.

Campbell, M., Holdsworth, C., Payne, T. and Dale, A. 1996: Sampling variance and design factors in the samples of anonymised records. *CCSR Occasional Paper No. 6*. Manchester: CCSR, University of Manchester.

Carter, R., Boudreau, J.-R, Briggs, M. 1991: Analysis of the risk of disclosure for census microdata. *Statistics Canada Working Paper*. Ottawa: Statistics Canada.

Centre for Longitudinal Studies 1999: *Longitudinal study newsletter*, No. 20. London: Institute of Education.

Charlton, J. 1998: Use of census sample of anonymised records (SARs) and survey data in combination to obtain estimates at local authority level. *Environment and Planning Series A* **30**, 775–784.

Cheng, Y. 1994: *Education and class: Chinese in Britain and the US*. Aldershot: Avebury.

Cheng, Y. 1996: The Chinese: upwardly mobile. In Peach, C. (ed.), *Ethnicity in the 1991 census, vol. 2: the ethnic minority populations of Great Britain*. London: HMSO, 161–180.

Choldin, H.M. 1994: *Looking for the last percent; the controversy over census undercounts*. New Brunswick: Rutgers University Press.

Clark, D.E., Knapp, T.A., White, N.E. 1996: Personal and location-specific characteristics and elderly interstate migration. *Growth and Change* **27**(3), 327–351.

Clegg, F. 1990: *Simple statistics*. Cambridge: Cambridge University Press.

Clifford, P. and Heath, A.F. 1993: The political consequences of social-mobility. *Journal of the Royal Statistical Society, Series A* **156**(Part 1), 51–61.

Coleman, D. and Salt, J. (eds), 1996: *Ethnicity in the 1991 census*, vol. 1. London: HMSO.

Collett, D. 1991: *Modelling binary data*. London: Chapman and Hall, ch. 6.

Cooke, T. and Bailey, A. 1996: Family migration and the employment of married women and men. *Economic Geography* **72**, 38–48.

Courgeau, D. 1973a: Migrants and migrations. *Population* **28**, 95-128.

Courgeau, D. 1973b: Migrations et découpages du territoire. *Population* **28**, 511–537.

Cullingford, D. and Openshaw, S. 1979: Deprived places or deprived people? *Discussion Paper 28*. Newcastle-upon-Tyne: Centre for Urban and Regional Development Studies.

Dale, A. 1993: Fieldwork and data processing. In Dale, A. and Marsh, C. (eds), *The 1991 census user's guide*. HMSO: London.

Dale, A. 1998: Confidentiality of official statistics: an excuse for privacy. In Dorling, D. and Simpson, S. (eds), *Statistics in society*. London: Arnold, 29-37.

Dale, A. and Davies, R. 1994: *Analyzing social and political change*, London: Sage.

Dale, A. and Egerton, M. 1997: Highly educated women: evidence from the National Child Development Study, *Research Studies RS25*. DfEE, London: The Stationery Office.

Dale, A. and Elliot, M. 1999: *Proposals for 2001 SARs: an assessment of disclosure risk*. Manchester: CCSR, Manchester University.

Dale, A. and Glover, J. 1990: An international comparison of women's employment patterns in the UK, France and the USA. *Department of Employment Research Paper No. 75*. London: Department of Employment.

Dale, A. and Holdsworth, C. 1997: Issues in the analysis of ethnicity in the 1991 British census. *Racial and Ethnic Studies* **20**, 160-181.

Dale, A. and Marsh, C. 1993: *The 1991 census user's guide*. London: HMSO.

Dale, A. and Openshaw, S. 1997: Adding area-based classifications to the samples of anonymised records (SAR) from the 1991 census, SARs newsletter. CCSR web site: http://les.man.ac.uk/ccsr/ccsrnew/areaclas.htm and *SARs Newsletter No. 12*. Manchester: CCSR, Manchester University.

Dale, A., Williams, M. and Dodgeon, B. 1996: *Housing deprivation and social change*. London: HMSO.

Demaris, A. 1992: *Logit modelling*. Sage University Paper 86. California: Sage.

Department of the Environment 1983: *Urban Deprivation Information Note 2*. London: Inner Cities Directorate, HMSO.

Department of the Environment 1992: *Inner cities research programme. Developing indicators to assess the potential for urban regeneration*. London: HMSO.

de Vaus, D.A. 1991: *Surveys in social research*. London: UCL Press.

Diamond, I., Clements, S., Stone, N. and Ingham, R. 1999: Spatial variation in teenage conceptions in south and west England. *Journal of the Royal Statistical Society, Series A* **162**(part 3), 273–289.

Dodd, T. 1987: A further investigation into the coverage of the postcode address file. In *Survey Methods Bulletin*, No. 21. London: Office for Population Censuses and Surveys, 35–40.

Dorling, D. and Atkins, D.J. 1995: Population density, change and concentration in Great Britain 1971, 1981, 1991. *Studies in Medical and Population Subjects*, No. 58. London: HMSO.

Dressler, W.W. 1994: Social status & health of families. *Social Science and Medicine* **39**, 1605–1613.

Drew, D., Gray, J. and Sporton, D. 1997: Ethnic differences in the educational participation of 16–19 year olds. In Karn, V. (ed.), *Ethnicity in the 1991 census: employment, education and housing among the ethnic minority populations of Britain*. London: HMSO, 17–28.

Duane-Richard, A.-M. 1998: How does the 'societal effect' shape the use of part-time work in France, the UK and Sweden? In O'Reilly, J. and Fagan, C. (eds), *Part-time prospects*. London: Routledge.

Duncan, G.T., Jabine, T.B. and de Wolf, V.A. 1993: *Private lives and public policies: confidentiality and accessibility of government statistics*. Washington, D.C.: National Academy Press.

Ebrahim, S. 1996. Principles of epidemiology in old age. In Ebrahim, S. and Kalache, A. (eds), *Epidemiology in old age*. London: BMJ Publishing Group.

Egerton, M. 1993: *The cultural capital hypothesis: a comparison between Hungary and the USA*. Report to the Spencer Foundation.

Eggebeen, D.J., Snyder, A.R. and Manning, W.D. 1996: Children in single-father families in demographic perspective. *Journal of Family Issues* **17**(4), 441–465.

Elliot, M. and Dale, A. 1999: Scenarios of attack: a data intruder's perspective on statistical disclosure risk. *Netherlands Official Statistics* **14**, 6–10.

Elliot, M. J., Skinner, C. J. and Dale, A. 1998: Special uniques, random uniques and sticky populations: some counterintuitive effects of geographical detail on disclosure risk. *Research in Official Statistics* **1**(2), 53–68.

Erikson, R. 1984: Social class of men, women and families. *Sociology* **18**, 500–514.

Erikson, R. and Goldthorpe, J.H. 1987: Commonality and variation in social fluidity in industrial nations. Part 1: A model for evaluating the 'FJH hypothesis'. *European Sociological Review* **3**(1), 54–77.

Erikson, R. and Goldthorpe, J.H. 1992: The CASMIN project and the American dream. *European Sociological Review* **8**(3), 283–305.

Erikson, R. and Goldthorpe, J. H. 1993: *The constant flux: a study of class mobility in industrial societies*. Oxford: Clarendon Press.

Erickson, B. and Nosanchuk, K. 1979: *Understanding data*. Milton Keynes: Open University Press.

Fang, D. and Brown, D. 1999: Geographic mobility of the foreign-born Chinese in large metropolises, 1985–1990. *International Migration Review* **33**(1), 137–155.

Featherman, D., Jones, F. and Hauser, R. 1975: Assumptions of social mobility research in the US: the case of occupational status. *Social Science Research* **4**, 329-60.

Fieldhouse, E. and Gould, M.I. 1998: Ethnic minority unemployment and local labour market conditions in Great Britain. *Environment & Planning A* **30**(5), 833–853.

Fieldhouse, E. and Tranmer, M. 1999: *Spatial mismatch or residualisation? Exploring labour market and neighbourhood variations in unemployment risk using census microdata.* Manchester: CCSR, Manchester University.

Fieldhouse, E.A. and Tye, R. 1996: Deprived people or deprived places? Exploring the ecological fallacy in studies of deprivation using the samples of anonymised records. *Environment & Planning A* **128**, 237–259.

Forrest, R., Gordon, D. 1993: *People and places: a 1991 census atlas of England.* Bristol: SAUS.

Fotheringham, A.S. and Pellegrini, P.A. 1996: Microdata for migration analysis: a comparison of sources in the US, Britain and Canada. *Area* **28**, 347–357.

Fox, J. 1991: *Regression diagnostics.* Quantitative Applications in the Social Sciences. London: Sage.

Fox, J. and Goldblatt, P.O. 1982: 1971–1975 longitudinal study of socio-demographic mortality differentials. *LS Series 1.* London: HMSO.

Ganzeboom, H., DeGraaf, P. and Treiman, D. 1992: A standard socio-economic index of occupational status. *Social Science Research* **21**, 1–56.

Ganzeboom, H. and Treiman, D. 1996: Internationally comparable measures of occupational status for the 1988 International Standard Classification of Occupations. *Social Science Research* **25**, 201–239.

Garonna, P. 1994: Statistical 'decentramento' in Italy: significance and implications. *Statistical Journal of the United Nations ECE,* **11**, 75–89.

Ghosh, M. and Rao, J.N.K. 1994: Small area estimation: an appraisal. *Statistical Science* **9**, 55–93.

Gilbert, N. 1993: *Analysing tabular data.* London: UCL Press.

Glaser, K. and Grundy, E. 1998: Migration and household change in the population aged 65 and over, 1971–1991. *International Journal of Population Geography* **4**, 1–17.

Glaser, K., Grundy, E. and Lynch, K. 1997a: Household transitions: coding independent and supported households among older persons. *Update – News from the LS User Group, No. 18,* 5–9.

Glaser, K., Murphy, M. and Grundy, E. 1997b: Limiting long-term illness and household structure among people aged 45 and over, Great Britain 1991. *Ageing and Society* **17**, 3–19.

Glover, J. 1992: Studying working women cross-nationally. *Work, Employment and Society* **6**(3): 489–498.

Goldman, N., Korenman, S. and Weinstein, R. 1995: Marital status and health among the elderly. *Social Science and Medicine* **40**(12), 1717–1730.

Goldstein, H. 1991: Nonlinear multilevel models: with an application to discrete response data. *Biometrika* **78**, 45.

Goldstein, H. 1994: Multilevel cross-classified models. *Sociological Methods and Research* **22**, 364.

Goldstein, H., 1995: *Multilevel statistical models,* 2nd edn. London: Arnold.

Goldstein, H., Rasbash, J., Plewis, I. *et al.* 1998: *A user's guide to MlwiN.* London: Institute of Education.

Goldthorpe, J. 1983: Women and class analysis: in defence of the conventional view. *Sociology* **17**, 465–488.

Goldthorpe, J.H. 1987. *Social mobility and class structure in modern Britain,* 2nd edn. Oxford: Clarendon Press.

Goldthorpe, J. and Payne, C. 1986: On the class mobility of women: results from different approaches to the analysis of recent British data. *Sociology* **20**, 531–555.

Gordon, D. 1995: Census-based deprivation indices: their weighting and validation. *Journal of Epidemiology and Community Health* **49**(Suppl. 2), S39–44.

Gordon, D. and Pantazis, C. (eds), 1997: *Breadline Britain in the 1990s.* Aldershot: Brookfield.

Gould, M.I. and Fieldhouse, E. 1997: Using the 1991 census SAR in a multilevel analysis of male unemployment. *Environment and Planning A* **29**, 611–628.

Gould, M.I. and Jones, K. 1996: Analysing perceived limiting long-term illness using UK census microdata. *Social Science and Medicine* **42**, 857–869.

Greenburg, B. and Voshell, L. 1990: The geographic component of disclosure risk for microdata. *SRD Research Report Census/SRD/RR-90/13.* Washington, DC: American Bureau of the Census.

Grundy, E. 1992a: The epidemiology of aging. In Brocklehurst, J.C., Tallis, R. and Fillet, H. (eds), *Textbook of gerontology and geriatric medicine.* Edinburgh: Churchill-Livingstone.

Grundy, E. 1992b: Socio-demographic variations in rates of movement into institutions among elderly people in England and Wales: an analysis of linked census and mortality data 1971–1985. *Population Studies* **46**, 65–84.

Grundy, E. and Glaser, K. 1997: Trends in, and transitions to, institutional residence among older people in England and Wales, 1971 to 1991. *Journal of Epidemiology and Community Health* **51**(5), 531–540.

Hakim, C. 1992: Explaining trends in occupational segregation: the measurement, causes and consequences of the sexual division of labour. *European Sociological Review* **8**, 127–152.

Hakim, C. 1993: Segregated and integrated occupations: a new approach to analysing social change. *European Sociological Review* **9**, 289–314.

Hakim, C. 1994: A century of change in occupational segregation 1891–1991. *Journal of Historical Sociology* **7**(4), 435–454.

Hakim, C. 1998: *Social change and innovation in the labour market.* Oxford: Oxford University Press.

Hamnett, C. 1979: Area-based explanations: a critical appraisal. In Herbert, D.T. and Smith, D.M. (eds), *Social problems and the city: geographical perspectives.* Oxford: Oxford University Press.

Harris, J. 2000: Web based access to area statistics. In Rees, P., Martin, D. and Williamson, P. (eds), *Census data system: resources, tools and developments.* London: The Stationery Office.

Harrop, A. and Plewis, I. 1993: A comparison of family structures in 1981: evidence from the General Household Survey and the OPCS Longitudinal Study. *Update – News from the LS User Group, No. 4.* London: SSRU, 11–16.

Hatch, S. and Sherrot, R. 1973: Positive discrimination and the distribution of deprivations. *Policy and Politics* **1**(3), 223–240.

Hattersely, L. and Creeser, R. 1995: *The OPCS Longitudinal Study.* London: HMSO.

Hauser, R.M. and Featherman, D.L. 1977: *The process of stratification: trends and analysis.* New York: Academic Press.

Hayes, L. and Al-Hamad, A. 1999: Residential change: differences in the movements and living arrangements of divorced men and women. In Boyle, P.J. and Halfacree, K.H. (eds), *Migration and gender in developed countries.* London: Routledge, 261–279.

Heady, P., Smith, S. and Avery, V. 1994: *1991 census: validation and coverage report.* London: HMSO.

Heady, P., Smith, S. and Avery, V. 1996: *1991 census: validation survey: quality report.* London: HMSO.

Heath, A. and McMahon, D. 1997: Education and occupational attainments: the impact of ethnic origins. In Karn, V. (ed.), *Ethnicity in the 1991 census: employment, education and housing among the ethnic minority populations of Britain.* London: HMSO, 91–113.

Heath, S. 1994: *A user guide to the SARs,* 2nd edn. Manchester: Census Microdata Unit, University of Manchester.

Heath, S. and Dale, A. 1994: Household & family formation in GB: the ethnic dimension. *Population Trends* **77**, 5–13.

Heath, S. and Miret Gamundi, P. 1996: Living in and out of the parental home in Spain and Great Britain: a comparative approach. *Working Paper Series, Number 2*. Cambridge: Cambridge Group for the History of Population and Social Structure.

Heckman, J.J. 1979: New evidence on the dynamics of female labor supply. In Lloyd, C.B., Andrews, E. and Gilroy, C. (eds), *Women in the Labour Market*. New York: Columbia University Press.

Hirschfield, A., 1994: Using the 1991 population census to study deprivation. *Planning Practice and Research* **9**(1).

HM Government 1999: *The 2001 Census of Population* (presented to Parliament by the Economic Secretary to the Treasury, the Secretary of State for Scotland, and the Secretary of State for Northern Ireland by Command of Her Majesty). Cm. 4253. London: The Stationery Office.

Hogan, H. 1993: The 1990 post-enumeration survey: operations and results. *Journal of the American Statistical Association* **88**, 1047–1060.

Holdsworth, C. 1995: Minimal household units. *SARs Newsletter No. 5*. Manchester: Census Microdata Unit, University of Manchester.

Holdsworth, C. 2000: Leaving home in Britain and Spain. *European Sociological Review* **16** (in press).

Holdsworth, C. and Dale, A. 1995: Ethnic homogeneity and family formation: evidence from the 1991 Household SAR. *CCSR Occasional Paper, No. 7*. Manchester: CCSR.

Holterman, S. 1975: Areas of urban deprivation in Great Britain: an analysis of census data. *Social Trends* **6**. London: HMSO.

Houseman, S. and Osawa, M. 1998: What is the nature of part-time work in the United States and Japan? In O'Reilly, J. and Fagan, C. (eds), *Part-time prospects*. London: Routledge, 232–251.

Humphrey, C. 1991: *The 1991 public use microdata files: a perspective from users*. Alberta: University Computing Systems, University of Alberta.

Illsley, R. 1955: Social class selection and class differences in relation to stillbirths and infant deaths. *British Medical Journal* **2**, 1520–1524.

Illsley, R. 1986: Occupational class, selection and the production of inequalities in health. *Quarterly Journal of Social Affairs* **2**, 151–165.

Index 99 Team (1999) *Report for formal consultation, stage 2: methodology for an index of multiple deprivation*. Oxford: University of Oxford.

International Labour Office 1990: *International Standard Classification of Occupations 1988*. Geneva: International Labour Office Publications.

James, A.D. 1998: What's love got to do with it?: economic viability and the likelihood of marriage among African American men. *Journal of Comparative Family Studies* **29**(2), 373–388.

Jasso, G. and Rosenzweig, M. 1990: *The new chosen people: immigrants in the US*. New York: Russell Sage Foundation.

Jones, K. 1994: Using multi-level modelling with area level data in the longitudinal study. In Creeser, R. (ed.), *LS user guide on analysis issues in the OPCS Longitudinal Study*. London: SSRU, City University.

Jones, K. 1997: Multilevel approaches to modelling contextuality: from nuisance to substance in the analysis of voting behaviour. In Dale, A. (ed.), *Exploiting national survey and census data: the role of locality and spatial effects*. CCSR Occasional Paper 12. Manchester: CCSR, Manchester University.

Jones, K. and Bullen, N. 1994: Contextual models of urban house prices: a comparison of fixed- and random-coefficient models developed by expansion. *Economic Geography* **70**, 252–272.

Jones, K. and Duncan, C. 1995: Individuals and their ecologies: analysing the geography of chronic illness within a multilevel modelling framework. *Health and Place* **1**, 27–40.

Jones, K. and Duncan, C. 1996: People and places: the multilevel model as general framework for the quantitative analysis of geographical data. In Longley, P. and Batty, M. (eds), *Spatial analysis: modelling in a GIS environment*. Cambridge: Cambridge GeoInformation, 79–104.

Jones, K., Gould, M.I. and Watt, R. 1998: Multiple contexts as cross-classified models: the Labour vote in the British General Election of 1992. *Geographical Analysis* **30**, 65–93.

Jonsson, J.O. and Mills, C. 1993: Social mobility in the 1970s and 1980s: a study of men and women in England and Sweden. *European Sociological Review* **9**, 229–247.

Joshi, H. and Paci, P. 1998: *Unequal pay for women and men*. Cambridge, MA: MIT Press.

Joshi, H., Paci, P. and Waldfogel, J. 1996: *The wages of motherhood: better or worse?* Welfare State Programme/122, STICERD. London: London School of Economics.

Keller-McNulty, S. and Unger, E.A. 1998: A database system prototype for remote access to information based on confidential data. *Journal of Official Statistics* **14**(4), 347–360.

Kerr, D. 1998: A review of procedures for estimating the net undercount of censuses in Canada, the United States, Britain and Australia. *Demographic Document No. 5*. Ottawa: Statistics Canada.

King, D. and Bolsdon, D. 1998: Adding policy value to household projections using the SARs. *Environment & Planning A* **30**, 867–880.

Kirby, A. 1981: Geographic contributions to the inner city deprivation debate: a critical assessment. *Area* **13**(3), 177–181.

Kitsul, P. and Philipov, D. 1981: The one-year/five year migration problem. In Rogers, A. (ed.), *Advances in multiregional mathematical demography, Research Report 81–6*. Laxenburg, Austria: International Institute for Applied Systems Analysis, 1–34.

Kralt, J. 1990: Ethnic origins in the Canadian census, 1871–1986. In Halli, S.S., Trovato, F. and Driedger, L. (eds), *Ethnic demography*. Ottawa: Carleton University Press, 13-29.

Krymkowski, D.H., Sawinski, Z. and Domanski, H. 1996: Classification schemes and the study of social mobility: a detailed examination of the Blau-Duncan categories. *Quality and Quantity* **30**(3), 301–321.

Ladipo, D. 1995: *Industrial change and social mobility: Black men in New York City and London 1970–1990*. The Jerome Levy Economics Institute, Bard College.

Langevin, Begeot, F. and Pearce, D. 1992: Censuses in the European Community. *Population Trends* **68**, 33–36.

Lathe, H. (ed.), 1995: *Labour and Income Dynamics Newsletter*, **4**(3), 6. Ottawa: Statistics Canada.

Lee, S.M. 1993: Racial classifications in the US census: 1890-1900. *Ethnic and Racial Studies* **16**, 75–94.

Leech, K. 1989: *A question in dispute: the debate about an 'ethnic' question in the census*. London: The Runnymede Trust.

Levitas, R. 1996: Fiddling while Britain burns? The 'measurement' of unemployment. In Levitas, R. and Guy, W. (eds), *Interpreting official statistics*. London: Routledge, 45–65.

Lewis, B. and Schnapper, D. (eds), 1994: *Muslims in Europe*. London: Pinter.

Lewis, J. and Townsend, A. (eds), 1989: *The north–south divide: regional change in Britain in the 1980s*. Liverpool: Paul Chapman.

Li, P. 1988: *Ethnic inequality in a class society*. Toronto: Wall and Thompson.

Lian, J.Z. and Matthews, D.R. 1998: Does the vertical mosaic still exist? Ethnicity and income in Canada, 1991. *Canadian Review of Sociology and Anthropology* **35**(4), 461–481.

Lin, G. 1997: Elderly migration: household versus individual approaches. *Papers in Regional Science* **76**(3), 285–300.

Long, L.H. 1988: *Migration and residential mobility in the United States.* Census Monograph Series: The Population of the United States in the 1980s. New York: Russell Sage Foundation.

Longford, N.T. 1998: Shrinkage estimation of local-area rates of economic activity. New methods for survey research 1988. *Proceedings of the Association for Survey Computing International Conference,* 21–22 August 1998, Southampton.

Longford, N.T. 1999: Multivariate shrinkage estimation of small area means and proportions, *Journal of the Royal Statistical Society, Series A* **162**(2), 227–245.

Lynn, P. and Lievesley, D. 1992: *Drawing general population samples in Great Britain.* London: SCPR.

Mack, J. and Lansley, S. 1985: *Poor Britain.* London: George Allen & Unwin.

Market Research Society 1991: *Occupational grouping: a job dictionary.* London: Market Research Society.

Marsh, C. 1988: *Exploring data: an introduction to data analysis for social scientists.* Cambridge: Polity Press, in association with Basil Blackwell.

Marsh, C. 1993a: An overview. In Dale, A. and Marsh, C. (eds), *The 1991 Census User's Guide.* London: HMSO.

Marsh, C. 1993b: The samples of anonymised records. In Dale, A. and Marsh, C. (eds), *The 1991 Census User's Guide.* London: HMSO.

Marsh, C. 1993c: The validation of census data. II. General issues. In Dale, A. and Marsh, C. (eds), *The 1991 Census User's Guide.* London: HMSO.

Marsh, C. 1993d: Privacy, confidentiality and anonymity in the 1991 census. In Dale, A. and Marsh, C. (eds), *The 1991 Census User's Guide.* London: HMSO.

Marsh, C., Dale, A. and Skinner, C. 1994: Safe data versus safe settings: access to microdata from the British census. *International Statistical Review* **62**(1): 35–53.

Marsh, C., Skinner, C., Arber, S. *et al.* 1991: The case for a sample of anonymized records from the 1991 census. *Journal of the Royal Statistical Society Series A* **154**, 305–340.

Marsh, C. and Teague, A. 1992: Samples of anonymised records from the 1991 Census. *Population Trends* **69**, 17–26.

Martin, D.F. and Taylor, J. 1995:, Enumerating the Aboriginal population of remote Australia: methodological and conceptual issues. *CAEPR Discussion Paper No. 91.* Canberra: Centre for Aboriginal Economic Policy Research, Australian National University.

Martin, J. and Roberts, C. 1984: *The women and employment survey.* London: HMSO.

Metcalf, D. and Nickell, S. J. 1982: Occupational mobility in Great Britain. In Ehrenberg, R.G. (ed.), *Research in labor economics.* Greenwich, CT: JAI Press.

Mills, I. and Teague, A. 1991: Editing and imputing data in the 1991 census. *Population Trends* **64**, 30–37.

Miret, P. 1995: Living together in Great Britain – displaying household structure through demographic pyramids. *Population Trends* **81**, 37–39.

Model, S. 1997: An occupational tale of two cities: minorities in London and New York. *Demography* **34**, 539–550.

Moser, C. and Kalton, G. 1971: *Survey methods in social investigation.* London: Heinemann Educational Books.

Muller, W., Blien, U. and Wirth, H. 1992: Disclosure risks of anonymous individual data. *Paper presented at the 1st International Seminar for Statistical Disclosure,* Dublin, September 1992.

Murphy, M., Glaser, K. and Grundy, E. 1997. Marital status and long-term illness in Great Britain. *Journal of Marriage and the Family* **59**, 156–164.

Murphy, M. and Wang, D. 1996: A dynamic multi-state projection model for making marital status population projections in England and Wales. In Dale, A. (ed.), *Exploiting national survey & census data.* CCSR Occasional Paper 10. Manchester: CCSR, University of Manchester.

Neave, H. 1981: *Elementary statistics tables*. London: Unwin Hyman.

Ní Bhrolcháin, M. 1990: The ethnicity question for the 1991 census; background and issues. *Racial and Ethnic Studies* **13**, 542–567.

Nickell, S.J. 1982: The determinants of occupational success in Britain. *Review of Economic Studies* **49**, 43–53.

Norment, L. 1995: Am I black, white or in between? Is there a plot to create a 'colored' buffer race in America? *Ebony* **50**, 108–111.

Office for National Statistics 1996a: *1991 census quality check report*. London: ONS.

Office for National Statistics 1996b: *Framework document*. London: ONS.

Office of Population Censuses and Surveys 1992: *Statistical information on population and housing 1996–2016: an invitation to shape the future*. London: OPCS.

Office of Population Censuses and Surveys 1993a: *Mid-1991 population estimates for England and Wales*. OPCS Monitor, PP1. London: OPCS.

Office of Population Censuses and Surveys 1993b: *1991 census. ethnic group and country of birth, Great Britain*, vol 1. London: HMSO.

Office of Population Censuses and Surveys/General Registrar Office (Scotland) (OPCS/GRO(S)) 1992: *Census definitions, Great Britain*. London: OPCS.

Office of Population Censuses and Surveys/General Registrar Office (Scotland) (OPCS/GRO(S)) 1994: *1991 Census User Guide 58: Under coverage in Great Britain*. London: OPCS.

Openshaw, S., Blake, M. and Wymer, C. 1994: *Using neurocomputing methods to classify Britain's residential areas. Working Paper 94/17*. Leeds: School of Geography, University of Leeds.

Openshaw, S., Evans, A., Duke-Williams, O. and Rees, P. 2000: A flexible outputs system for the 2001 census including a measure of confidentiality risks in census data. In Rees, P., Martin, D. and Williamson, P. (eds), *The census data system: resources, tools and developments*. London: The Stationery Office (in press).

O'Reilly, J. 1996: Theoretical considerations in cross-national employment research. *Sociological Review Online* **1**, 1 (http://www.soc.surrey.ac.uk).

Oswald, A.J. 1997: Happiness and economic performance. *Economic Journal* **107**, 1815–1832.

Pagano, R 1986: *Understanding statistics in the behavioural sciences*, 2nd edn. West Publishing: St Paul.

Payne, J. 1987: Does unemployment run in families? Some findings from the General Household Survey. *Sociology* **21**(2), 199–214.

Payne, C., Payne, J. and Heath, A. 1994: Modelling trends in multi-way tables. In Dale, A. and Davies, R. (eds), *Analyzing social change*. London: Sage.

Pearlmann, J. 1997: Multiracials, intermarriage, ethnicity. *Society* 34, 20–24.

Pfau-Effinger, B. 1998: Gender cultures and the gender arrangement – a theoretical framework for cross-national gender research, *Innovations* **11**(2), 147–165.

Plewis, I. 1994: Longitudinal multilevel models. In Dale, A. and Davies, R. (eds), *Analyzing social change*. London: Sage.

Plewis, I. 1997: *Statistics in education*. London: Arnold.

Power, C., Manor, O., Fox, J. and Fogelman, K. 1990: Health in childhood and social inequalities in health in young adults. *Journal of the Royal Statistical Society A* **153**(Part 1), 17–28.

Prandy, K. 1990: Revised Cambridge scale of occupation. *Sociology* 24(4), 629–655.

Prandy, K. 1992: Cambridge scale scores for CASOC groupings. *Working paper No. 11*. Cambridge: Social and Political Sciences.

Rachedi, N. 1994: Elites of Maghrebian extraction in France. In Lewis, B. and Schnapper, D. (eds), *Muslims in Europe*. London: Pinter.

Raftery, A.E. 1985: Social-mobility measures for cross-national comparisons. *Quality and Quantity* **19**(2), 167–182.

Raudenbush, S.W. 1993: A crossed random effects model for unbalanced data with applications in cross-sectional and longitudinal research. *Journal of Educational Statistics* **18**, 321–349.

Redfern, P. 1981: Census 1981 – an historical and international perspective. *Population Trends* **23**, 3–15.

Redfern, P. 1987: *A study of the future of the census of population: alternative approaches.* Luxembourg: Office for Official Publications of the European Communities.

Rees, P.H. 1977: The measurement of migration from census data and other sources. *Environment and Planning A* **9**, 247–272.

Rees, P.H. 1994: Estimating and projecting the populations of urban communities. *Environment and Planning Series A* **26**, 1671–1697.

Research International Labour Office 1990: *International Standard Classification of Occupations 1988.* Geneva: International Labour Office Publications.

Robinson, P. 1986: Women's occupational attainment. *Social Science Research* **15**, 323–346.

Robinson, W.S. 1950: Ecological correlations and the behaviour of individuals. *American Sociological Review* **15**, 351–357.

Rogerson, P.A. 1990: Migration analysis using data with time intervals of differing widths. *Papers of the Regional Science Association* **68**, 97–106.

Roos, P. and Hennessy, J. 1987: Assimilation or exclusion? Japanese and Mexican Americans in California. *Sociological Forum* **2**, 278-304.

Rose, D. and Sullivan, O. 1993: *Introducing data analysis for social sciences.* Buckingham: Open University Press.

Rowntree, D. 1991: *Statistics without tears.* London: Penguin.

Rubery, J., Fagan, C and Smith, M. 1998: *Women and European employment.* London: Routledge.

*SARs Newsletter* 1995: No. 6. Manchester: Census Microdata Unit.

*SARs Newsletter* 1998: No. 12. Manchester: Census Microdata Unit.

Scheuch, E.K. 1990: The development of comparative research: towards causal explanations. In Oyen, E. (ed.), *Comparative methodology.* Sage Studies in International Sociology 40. London: Sage, 19–37.

Scott, J. and Duncombe, J. 1991: A cross-national comparison of gender-role attitudes: is the working mother selfish? *Working Paper, ESRC Research Centre on Micro-social Change. Paper 9.* Colchester: University of Essex.

Shouls, S., Congdon, P. and Curtis, S. 1996: Modelling inequality in reported long term illness: combining individual and area characteristics. *Journal of Epidemiology and Community Health* **50**, 366–376.

Siltanen, J., Jarman, J. and Blackburn, R. 1995: *Gender inequality in the labour market: occupational concentration and segregation.* Geneva: International Labour Office.

Simpson, S. (ed.), 1993: *Census indicators of local poverty and deprivation: methodological issues.* London: Local Authorities Research and Intelligence Association (LARIA).

Simpson, S. 1996: Resource allocation by measures of relative social need in geographical areas: the relevance of the signed $\chi^2$, the percentage and the raw count. *Environment & Planning A* **28**(3), 537–554.

Simpson, S. 1999: Statistical exclusion and social exclusion: the impact of missing data. *Radical Statistics* **71**, 45-60.

Simpson, S., Cossey, R. and Diamond, I. 1997: 1991 population estimates for areas smaller than districts. *Population Trends* **90**, 31–39.

Simpson, S. and Dorling, D. 1994: Those missing millions: implications for social statistics of non response to the 1991 census. *Journal of Social Policy* **23**(4), 543–567.

Simpson, S. and Middleton, E. 1998: Characteristics of census undercount: what do we know already? Adjusting census output for undercount. In Simpson, S. (ed.), *A one number census: proceedings of a research workshop.* CCSR Occasional Paper 15. Manchester: University of Manchester, 1–14.

Simpson, S. and Middleton, E. 1999: Undercount of migration in the UK 1991 census and its impact on counterurbanisation and population projections. *International Journal of Population Geography* **5**, 387–405.

Skinner, C.J., Holt, D. and Smith, T.M.F. (eds), 1989: *Analysis of complex surveys*. Chichester: John Wiley.

Skinner, C.J., Marsh, C., Openshaw, S. and Wymer, C. 1994: Disclosure control for census microdata, *Journal of Official Statistics* **10**, 31–51.

Sloggett, A. and Joshi, H. 1994: Higher mortality in deprived areas: community or personal disadvantage? *British Medical Journal* **309**, 1470–1474.

Smith, D.M. 1977: *Human geography: a welfare approach*. London: Edward Arnold.

Smith, S. 1993: *Electoral registration in 1991*. London: HMSO.

Stanworth, M. 1984: Women and class analysis: a reply to Goldthorpe, *Sociology* **18**, 159–170.

Statistics Canada 1992a: *1991 census handbook* (cat. 92-305E). Ottawa: Supply and Services Canada.

Statistics Canada 1992b: *Immigration and citizenship: the nation. 1991 census of Canada*. Ottawa: Ministry of Industry, Science, and Technology.

Statistics Canada 1994: *User documentation for public use microdata file on individuals. 1991 census of Canada*. Ottawa: Ministry of Industry, Science and Technology.

Statistics Canada 1995: *Occupation according to the 1991 Standard Occupational Classification*. 1991 Census Technical Reports; Reference Products Series. Ottawa: Ministry of Industry, Science and Technology.

Statistics Canada 1998: 1996 census: ethnic origin, visible minorities. *The Daily*, 17 February.

Stewart, A., Prandy, K. and Blackburn, R.M. 1980: *Social stratification and occupations*. London: Macmillan.

Stewart, M.B. 1983: Racial discrimination and occupational attainment in Britain. *Economic Journal* **93**, 521–541.

Szakoiczai, A. and Fustos, L. 1998: Value systems in axial moments: a comparative analysis of 24 European countries. *European Sociological Review* **14**(3), 211–230.

Thomas, M. 1994: Income on the General Household Survey. *OPCS Survey Methodology Bulletin*, January, 19–24. London: OPCS.

Tienda, M. Smith, S. A. and Ortiz. V. 1987: Industrial restructuring, gender segregation, and sex differences in earnings. *American Sociological Review* **52**(2), 195–210.

Townsel, L. 1996: Neither Black nor White. *Ebony* **52**, 44–49.

Townsend, P., Phillimore, P. and Beattie, A. 1988: *Health and deprivation: inequality and the North*. London: Croom Helm.

Tranmer, M. and Steel, D.G. 1998: Using census data to investigate the causes of the ecological fallacy. *Environment and Planning Series A* **30**, 817–831.

Tukey, J.W. 1977: *Exploratory data analysis*. London: Addison-Wesley.

Turton, I. 1999: A pilot flexible output system. *Paper presented at ESRC 2001 Census Development Programme, First workshop: project proposals and feedback*, October 1999, London.

United Nations 1949: *Population census handbook 1949*. New York: United Nations Organization.

US Bureau of the Census 1990: Census of population and housing, 1990 United States: Public Use Microdata Sample: 1-percent sample (ICPSR 9951). Washington, DC: US Bureau of the Census.

US Dept of Commerce 1993: *PUMS*. Washington DC: Bureau of the Census.

Verbrugge, L.M. 1979: Marital status and health. *Journal of Marriage and the Family* **41**, 267–285.

Waldfogel, J. 1993: *Women working for less: a longitudinal analysis of the family gap*. Welfare State Programme, No. WSP/93, STICERD. London: London School of Economics.

Wall, R. 1996: Comparer ménages et familles au niveau Européen: problèmes et perspectives. *Population* **51**, 93–115.

Wallace, M. and Denham, C. 1996: *The ONS classification of wards. SMPS 60.* London: HMSO.

Wallace, M., Charlton, J. and Denham, C. 1995: The new OPCS area classification. *Population Trends* **79**, 15–30.

Waters, M.C. 1990: *Ethnic options.* Berkeley: University of California Press.

Watts, M. 1993: Explaining trends in occupational segregation: some comments. *European Sociological Review* **9**, 315–319.

Watts, M. 1994: A critique of marginal matching. *Work, Employment and Society* **8**, 421–431.

Watts, M. 1997: The measurement of occupational gender segregation. *Journal of the Royal Statistical Society, Series A* **160**(Part 1), 141–145.

Wiggins, R. 1993: The validation of census data. I. Post-enumeration survey approaches. In Dale, A. and Marsh, C. (eds), *The 1991 Census User's Guide.* London: HMSO.

Williamson, P., Birkin M. and Rees P.H. 1998: The estimation of population microdata by using data from small area statistics and samples of anonymised records. *Environment & Planning A* **30**, 785–816.

Wong, G. and Mason, W. 1984: The hierarchical logistic regression model for multi-level analysis. *Journal of the American Statistical Association* **80**, 513–524.

Wrigley, N. 1985: *Categorical data analysis for geographers and environmental scientists.* Harlow: Longman.

# Index

Page numbers in *italics* refer to tables and boxed material. Page numbers in **bold** refer to figures.

Lightning Source UK Ltd.
Milton Keynes UK
UKOW06f2153041017
310413UK00003B/178/P